STUDENT SOLUTIONS MANUAL
to accompany

Charles P. McKeague

Judy Barclay
Cuesta College

Saunders College Publishing

Harcourt College Publishers
Fort Worth   Philadelphia   San Diego   New York   Orlando   Austin
San Antonio   Toronto   Montreal   London   Sydney   Tokyo

Copyright © 1999 by Harcourt, Inc.

All rights reserved. No part of this publication may be reproduced or
transmitted in any form or by any means, electronic or mechanical,
including photocopy, recording, or any information storage and
retrieval system, without permission in writing from the publisher.

Requests for permission to make copies of any part of the work should
be mailed to: Permissions Department, Harcourt, Inc., 6277 Sea Harbor
Drive, Orlando, FL 32887-6777.

Portions of this work were published in previous editions.

Printed in the United States of America

ISBN 0-03-024971-6

9 0 1 2 3 4 5 6 7 8 202 10 9 8 7 6 5 4 3 2

Preface to the Student

This manual contains detailed explanations to every other odd exercise from the problem sets and every exercise from the chapter tests. There is no substitute for working through a problem on your own, but I hope that this manual will end some of the frustration involved in problem solving.

Try all exercises on your own before consulting this manual. It will also be helpful to study one problem and then use the same technique to solve a similar one whose solution is not included in this manual. The solution of a problem is the important part, not the answer.

You would not think about becoming a good tennis player without a great deal of practice. The same is true in mathematics. I wish you good luck in your problem solving.

When using a calculator to compute the answer to an exercise, do not round until the end of the calculation. In this manual, intermediate steps have been shown to three or four decimal places for you to check. However, the calculations have been performed on a calculator without rounding until the last step. In other words, the calculations only appear to be rounded, they are not.

Judy Barclay

Contents

CHAPTER 1 The Six Trigonometric Functions

Problem Set 1.1

1. 10° is an acute angle.

The complement of 10° is 80° because $10° + 80° = 90°$.

The supplement of 10° is 170° because $10° + 170° = 180°$.

5. 120° is an obtuse angle.

The complement of 120° is $-30°$ because $120° + (-30°) = 90°$.

The supplement of 120° is 60° because $120° + 60° = 180°$.

9.
$$
\begin{aligned}
\alpha &= 180° - (\angle A + \angle D) && \text{The sum of the angles of a triangle is } 180° \\
&= 180° - (30° + 90°) && \text{Substitute given values} \\
&= 180° - 120° && \text{Simplify} \\
&= 60°
\end{aligned}
$$

13.
$$
\begin{aligned}
\angle A &= 180° - (\alpha + \beta + \angle B) && \text{The sum of the angles of a triangle is } 180° \\
&= 180° - (100° + 30°) && \text{Substitute given values} \\
&= 180° - 130° && \text{Simplify} \\
&= 50°
\end{aligned}
$$

17.
$$
\begin{aligned}
\alpha + \beta &= 90° && \alpha \text{ and } \beta \text{ are complementary} \\
\beta &= 90° - \alpha && \text{Subtract } \alpha \text{ from both sides} \\
&= 90° - 25° && \text{Substitute given value} \\
&= 65° && \text{Simplify}
\end{aligned}
$$

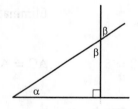

21. One complete revolution equals 360°.

In 4 hours, the hour hand revolves $\dfrac{4}{12}$ or $\dfrac{1}{3}$ of a revolution.

$\dfrac{1}{3}$ of 360° = 120°.

25.
$$c^2 = a^2 + b^2 \qquad \text{Pythagorean theorem}$$
$$= (4)^2 + (3)^2 \qquad \text{Substitute given values}$$
$$= 16 + 9 \qquad \text{Simplify}$$
$$= 25$$

Therefore, $c = \pm 5$. Our only solution is $c = 5$, because we cannot use $c = -5$.

29.
$$a^2 + b^2 = c^2 \qquad \text{Pythagorean theorem}$$
$$a^2 = c^2 - b^2 \qquad \text{Subtract } b^2 \text{ from both sides}$$
$$= (13)^2 - (12)^2 \qquad \text{Substitute given values}$$
$$= 169 - 144 \qquad \text{Simplify}$$
$$= 25$$

Therefore, $a = \pm 5$. Our only solution is $a = 5$, because we cannot use $a = -5$.

33.
$$x^2 = (2)^2 + (2\sqrt{3})^2 \qquad \text{Pythagorean theorem}$$
$$= 4 + 12 \qquad \text{Simplify}$$
$$= 16$$

Therefore, $x = \pm 4$. Our only solution is $x = 4$, because we cannot use $x = -4$.

37.
$$(BD)^2 = (CD)^2 + (BC)^2 \qquad \text{Pythagorean theorem}$$
$$5^2 = (CD)^2 + (4)^2 \qquad \text{Substitute given values}$$
$$25 = (CD)^2 + 16 \qquad \text{Simplify}$$
$$9 = (CD)^2 \qquad \text{Subtract 16 from both sides}$$
$$CD = 3 \;\; \text{or} \;\; CD = -3 \qquad \text{Take square root of both sides}$$
$$CD = 3 \qquad \text{Eliminate negative solution}$$

Therefore, $AC = 2 + 3 = 5 \qquad AC = AD + DC$

$$(AB)^2 = (AC)^2 + (BC)^2 \qquad \text{Pythagorean theorem}$$
$$= 5^2 + 4^2 \qquad \text{Substitute given values}$$
$$= 25 + 16 \qquad \text{Simplify}$$
$$= \sqrt{41}$$
$$AB = \sqrt{41} \;\; \text{or} \;\; AB = -\sqrt{41} \quad \text{Take square root of both sides}$$
$$AB = \sqrt{41} \qquad \text{Eliminate negative solution}$$

41. This is an isosceles triangle. Therefore, the altitude must bisect the base.

$$x^2 = (18)^2 + (13.5)^2 \qquad \text{Pythagorean theorem}$$
$$= 324 + 182.5 \qquad \text{Simplify}$$
$$= 506.25$$
$$x = 22.5 \quad \text{or} \quad x = -22.5 \qquad \text{Take square root of both sides}$$
$$x = 22.5 \text{ feet} \qquad \text{Eliminate negative solution}$$

45. The longest side is 8 which is twice the shortest side.

Therefore, the shortest side is 4.

The side opposite the 60° angle is $4\sqrt{3}$.

49. The shortest side is 20 feet

The longest side is twice the shortest side.

Therefore, $x = 2(20)$
$$x = 40 \text{ feet}$$

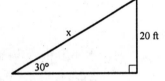

53. $\text{hypotenuse} = \dfrac{4}{5} \cdot \sqrt{2} \qquad \text{Hypotenuse is } t \cdot \sqrt{2}$

$$= \dfrac{4\sqrt{2}}{5} \qquad \text{Simplify}$$

57. $\text{hypotenuse} = t\sqrt{2} \qquad t \text{ is the shorter side}$

$$4 = t\sqrt{2} \qquad \text{Substitute given value}$$

$$\dfrac{4}{\sqrt{2}} = t \qquad \text{Divide both sides by } \sqrt{2}$$

$$t = 2\sqrt{2} \qquad \text{Rationalize denominator by multiplying numerator and}$$
$$\text{denominator by } \sqrt{2}$$

61. (a) $\text{hypotenuse} = t\sqrt{2} \qquad t \text{ is the edge of the cube}$

$$= 1\sqrt{2} \qquad \text{Substitute given value}$$

$$= \sqrt{2} \qquad \text{Simplify}$$

Therefore, $\text{CH} = \sqrt{2}$

This problem is continued on the next page.

(b)
$$(CF)^2 = (CH)^2 + (FH)^2 \qquad \text{Pythagorean theorem}$$
$$= (\sqrt{2})^2 + (1)^2 \qquad \text{Substitute given values}$$
$$= 2 + 1 \qquad \text{Simplify}$$
$$= 3$$
$$CF = \pm\sqrt{3} \qquad \text{Take square root of both sides}$$
$$CF = \sqrt{3} \qquad \text{Eliminate the negative solution}$$

65.
$$(CF)^2 = (CH)^2 + (FH)^2 \qquad \text{Pythagorean theorem}$$
$$= (t\sqrt{2})^2 + (t)^2 \qquad \text{From Problem 61a above}$$
$$= 2t^2 + t^2 \qquad \text{Simplify}$$
$$= 3t^2$$

$$(3)^2 = 3t^2 \qquad \text{Substitute given value}$$
$$3 = t^2 \qquad \text{Divide both sides by 3}$$
$$t = \pm\sqrt{3} \qquad \text{Take square root of both sides}$$
$$t = \sqrt{3} \qquad \text{Eliminate the negative solution}$$

Problem Set 1.2

13. If we let $x = 0$, the equation $3x + 2y = 6$ becomes:
$$3(0) + 2y = 6$$
$$2y = 6$$
$$y = 6$$

This gives us $(0, 3)$ as one solution to $3x + 2y = 6$.

If we let $y = 0$, the equation $3x + 2y = 6$ becomes:
$$3x + 2(0) = 6$$
$$3x = 6$$
$$x = 2$$

This gives us $(2, 0)$ as a second solution to $3x + 2y = 6$.

Graphing the points $(0, 3)$ and $(2, 0)$ and then drawing a line through them, we have the graph of $3x + 2y = 6$.

17. The vertex of this parabola is at $(0, -4)$.

If we let $x = -2$, the equation $y = x^2 - 4$ becomes $y = (-2)^2 - 4 = 4 - 4 = 0$.
This gives us $(-2, 0)$ as a point on the curve.

If we let $x = -1$, the equation $y = x^2 - 4$ becomes $y = (-1)^2 - 4 = 1 - 4 = -3$.
This gives us $(-1, -3)$ as a point on the curve.

If we let $x = 1$, the equation $y = x^2 - 4$ becomes $y = 1^2 - 4 = 1 - 4 = -3$.
This gives us $(1, -3)$ as a point on the curve.

If we let $x = 2$, the equation $y = x^2 - 4$ becomes $y = 2^2 - 4 = 4 - 4 = 0$
This gives us $(2, 0)$ as a point on the curve.

Graphing the points $(-2, 0)$, $(-1, -3)$, $(0, -4)$, $(1, -3)$, and $(2, 0)$, and then drawing a smooth curve through them, we have the graph of the parabola $y = x^2 - 4$.

21. The center of this circle is $(0, 0)$ and the radius is 5.

25. From the graph of $x^2 + y^2 = 25$, we can see that $(0, 5)$ and $(5, 0)$ are the points at which the line $x + y = 5$ will intersect the circle.

29.
$$r = \sqrt{(x_2 - x_1)^2 + (y_2 - y_1)^2} \qquad \text{Distance formula}$$
$$= \sqrt{(3 - 6)^2 + (7 - 3)^2} \qquad \text{Substitute given values}$$
$$= \sqrt{(-3)^2 + 4^2} \qquad \text{Simplify}$$
$$= \sqrt{9 + 16}$$
$$= \sqrt{25} = 5$$

33.
$$r = \sqrt{(x_2 - x_1)^2 + (y_2 - y_1)^2} \qquad \text{Distance formula}$$
$$= \sqrt{[3 - (-2)]^2 + (-5 - 1)^2} \qquad \text{Substitute given values}$$
$$= \sqrt{5^2 + (-6)^2} \qquad \text{Simplify}$$
$$= \sqrt{25 + 36}$$
$$= \sqrt{61}$$

37.

$$r = \sqrt{(x_2 - x_1)^2 + (y_2 - y_1)^2} \qquad \text{Distance formula}$$

$$= \sqrt{(3 - 0)^2 + (-4 - 0)^2} \qquad \text{Substitute given values}$$

$$= \sqrt{3^2 + (-4)^2} \qquad \text{Simplify}$$

$$= \sqrt{9 + 16}$$

$$= \sqrt{25} = 5$$

41. First, we convert 1.2 miles to feet: $1.2 \text{ mi} = 1.2(5{,}280) \text{ ft} = 6{,}336 \text{ ft}$

$$c^2 = a^2 + b^2 \qquad \text{Pythagorean theorem}$$

$$= (2{,}640)^2 + (6{,}336)^2 \qquad \text{Substitute given values}$$

$$= 6{,}969{,}600 + 40{,}144{,}896 \quad \text{Simplify}$$

$$= 47{,}114{,}496$$

$$c = 6{,}864 \text{ ft (or 1.3 mi)} \qquad \text{Take square root of both sides}$$

45. Quadrants II and III lie to the left of the y-axis. Therefore, all points in these two quadrants have negative x-coordinates.

49. The cannonball is at the ground $(y = 0)$ when $x = 0$ and when $x = 160$. The x-coordinate of the vertex (the maximum) will be at $\frac{1}{2}(160)$ or 80, and the y-coordinate will be 60.

We can now sketch the parabola through the points $(0, 0)$, $(80, 60)$, and $(160, 0)$.

The equation will be in the form $y = a(x - 80)^2 + 60$. We will use the point $(160, 0)$ to find a.

Let $x = 160$ and $y = 0$:

$$y = a(x - 80)^2 + 60$$
$$0 = a(160 - 80)^2 + 60$$
$$-60 = a(80)^2$$
$$-60 = 6400a$$
$$a = -\frac{3}{320}$$

Therefore, the equation is $y = -\frac{3}{320}(x - 80)^2 + 60$.

53. The complement of a 60° angle is $90° - 60° = 30°$.

57. The supplement of a 90° angle is $180° - 90° = 90°$.

61. An angle coterminal with an angle of $-210°$ is $-210° + 360° = 150°$.

65. **a.** If we draw 225° in standard position, we see that the terminal side is along the line $y = x$. Since the terminal side lies in the third quadrant, x and y are both negative. Some of the points on the terminal sides are:

$(-1, -1), (-3, -3), (-\sqrt{2}, -\sqrt{2})$, and $(-\frac{1}{2}, -\frac{1}{2})$.

b. To find the distance from $(0, 0)$ to $(-3, -3)$, we use the distance formula:

$$r = \sqrt{(-3 - 0)^2 + (-3 - 0)^2}$$
$$= \sqrt{(-3)^2 + (-3)^2}$$
$$= \sqrt{9 + 9}$$
$$= \sqrt{18} \text{ or } 3\sqrt{2}$$

c. The angle between 0° and $-360°$ that is coterminal with 225° is $-135°$.

69. **a.** If we draw $-45°$ in standard position, we see that the terminal side is along the line $y = -x$. Since the terminal side lies in the fourth quadrant, x is positive and y is negative. Some of the points on the terminal side are: $(1, -1), (3, -3), (\sqrt{2}, -\sqrt{2})$, and $(\frac{1}{2}, -\frac{1}{2})$.

b. To find the distance from $(0, 0)$ to $(3, -3)$, we use the distance formula:

$$r = \sqrt{(3 - 0)^2 + (-3 - 0)^2}$$
$$= \sqrt{3^2 + (-3)^2}$$
$$= \sqrt{9 + 9}$$
$$= \sqrt{18} \text{ or } 3\sqrt{2}$$

c. The angle between 0° and 360° that is coterminal with $-45°$ is 315°.

73. $r = \sqrt{(3 - 0)^2 + (-2 - 0)^2}$

$$= \sqrt{3^2 + (-2)^2}$$
$$= \sqrt{9 + 4}$$
$$= \sqrt{13}$$

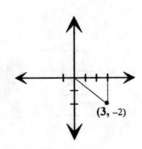

(3, –2)

Problem Set 1.3

1. $(x, y) = (3, 4)$

$\sin \theta = \dfrac{y}{r} = \dfrac{4}{5}$

$x = 3$ and $y = 4$

$\cos \theta = \dfrac{x}{r} = \dfrac{3}{5}$

$r = \sqrt{3^2 + 4^2}$

$\tan \theta = \dfrac{y}{x} = \dfrac{4}{3}$

$\quad = \sqrt{9 + 16}$

$\cot \theta = \dfrac{x}{y} = \dfrac{3}{4}$

$\quad = \sqrt{25}$

$\sec \theta = \dfrac{r}{x} = \dfrac{5}{3}$

$\quad = 5$

$\csc \theta = \dfrac{r}{y} = \dfrac{5}{4}$

5. $(x, y) = (-5, 12)$

$\sin \theta = \dfrac{y}{r} = \dfrac{12}{13}$

$x = -5$ and $y = 12$

$\cos \theta = \dfrac{x}{r} = -\dfrac{5}{13}$

$r = \sqrt{(-5)^2 + (12)^2}$

$\tan \theta = \dfrac{y}{x} = -\dfrac{12}{5}$

$\quad = \sqrt{25 + 144}$

$\cot \theta = \dfrac{x}{y} = -\dfrac{5}{12}$

$\quad = \sqrt{169}$

$\sec \theta = \dfrac{r}{x} = -\dfrac{13}{5}$

$\quad = 13$

$\csc \theta = \dfrac{r}{y} = \dfrac{13}{12}$

9. $(x, y) = (a, b)$

$\sin \theta = \dfrac{y}{r} = \dfrac{b}{\sqrt{a^2 + b^2}}$

$x = a$ and $y = b$

$\cos \theta = \dfrac{x}{r} = \dfrac{a}{\sqrt{a^2 + b^2}}$

$r = \sqrt{a^2 + b^2}$

$\tan \theta = \dfrac{y}{x} = \dfrac{b}{a}$

$\cot \theta = \dfrac{x}{y} = \dfrac{a}{b}$

$\sec \theta = \dfrac{r}{x} = \dfrac{\sqrt{a^2 + b^2}}{a}$

$\csc \theta = \dfrac{r}{y} = \dfrac{\sqrt{a^2 + b^2}}{b}$

13. $(x, y) = (\sqrt{3}, -1)$ $\sin\theta = \dfrac{y}{r} = -\dfrac{1}{2}$

$x = \sqrt{3}$ and $y = -1$ $\cos\theta = \dfrac{x}{r} = \dfrac{\sqrt{3}}{2}$

$r = \sqrt{(\sqrt{3})^2 + (-1)^2}$ $\tan\theta = \dfrac{y}{x} = -\dfrac{1}{\sqrt{3}}$

$= \sqrt{3+1}$ $\cot\theta = \dfrac{x}{y} = -\dfrac{\sqrt{3}}{1} = -\sqrt{3}$

$= \sqrt{4}$ $\sec\theta = \dfrac{r}{x} = \dfrac{2}{\sqrt{3}}$

$= 2$ $\csc\theta = \dfrac{r}{y} = -\dfrac{2}{1} = -2$

17. $(x, y) = (60, 80)$ $\sin\theta = \dfrac{y}{r} = \dfrac{80}{100} = \dfrac{4}{5}$

$x = 60$ and $y = 80$ $\cos\theta = \dfrac{x}{r} = \dfrac{60}{100} = \dfrac{3}{5}$

$r = \sqrt{(60)^2 + (80)^2}$ $\tan\theta = \dfrac{y}{x} = \dfrac{80}{60} = \dfrac{4}{3}$

$= \sqrt{3600 + 6400}$ $\cot\theta = \dfrac{x}{y} = \dfrac{60}{80} = \dfrac{3}{4}$

$= \sqrt{10{,}000}$ $\sec\theta = \dfrac{r}{x} = \dfrac{100}{60} = \dfrac{5}{3}$

$= 100$ $\csc\theta = \dfrac{r}{y} = \dfrac{100}{80} = \dfrac{5}{4}$

21. $(3, 4)$ lies on the terminal side of θ. Therefore,

$x = 3, \ y = 4,$ and $r = \sqrt{3^2 + 4^2} = \sqrt{9 + 16} = \sqrt{25} = 5$

$\sin\theta = \dfrac{y}{r} = \dfrac{4}{5}$

$\cos\theta = \dfrac{x}{r} = \dfrac{3}{5}$

$\tan\theta = \dfrac{y}{x} = \dfrac{4}{3}$

25. $(x, y) = (9.36, 7.02)$ 　　　　 $\sin\theta = \dfrac{y}{r} = \dfrac{7.02}{11.7} = 0.6$

$x = 9.36$ and $y = 7.02$

$r = \sqrt{(9.36)^2 + (7.02)^2}$ 　　 $\cos\theta = \dfrac{x}{r} = \dfrac{9.36}{11.7} = 0.8$

$ = \sqrt{87.6096 + 49.2804}$

$ = \sqrt{136.89}$

$ = 11.7$

29. A point on the terminal side of an angle of 90° is (0, 1).

$(x, y) = (0, 1)$ 　　　　 $\sin\theta = \dfrac{y}{r} = \dfrac{1}{1} = 1$

$x = 0$ and $y = 1$ 　　　 $\cos\theta = \dfrac{x}{r} = \dfrac{0}{1} = 0$

$r = \sqrt{0^2 + 1^2}$ 　　　 $\tan\theta = \dfrac{y}{x} = \dfrac{1}{0}$ (undefined)

$ = \sqrt{1}$

$ = 1$

33. A point on the terminal side of an angle of 0° is (1, 0).

$(x, y) = (1, 0)$ 　　　　 $\sin\theta = \dfrac{y}{r} = \dfrac{0}{1} = 0$

$x = 1$ and $y = 0$ 　　　 $\cos\theta = \dfrac{y}{r} = \dfrac{1}{1} = 1$

$r = \sqrt{1^2 + 0^2}$ 　　　 $\tan\theta = \dfrac{y}{x} = \dfrac{0}{1} = 0$

$ = \sqrt{1}$

$ = 1$

37. $\sin\theta = \dfrac{y}{r}$ 　 (r is always positive)

If $\sin\theta$ is negative, then y must be negative.

y is negative in quadrants III and IV.

Therefore, $\sin\theta$ is negative in QIII and QIV.

41. The sine function is positive in quadrants I and II.

The tangent function is positive in quadrants I and III.

Therefore, both functions are positive in quadrant I.

The sine function is negative in quadrants III and IV.

The tangent function is negative in quadrants II and IV.

Therefore, both functions are negative in quadrant IV.

43. $\sin \theta = \dfrac{y}{r} = \dfrac{12}{13}$ and θ terminates in QI.

$y = 12$ and $r = 13$

$$x^2 + y^2 = r^2$$

$$x^2 + (12)^2 = (13)^2$$

$$x^2 + 144 = 169$$

$$x^2 = 25$$

$$x = \pm 5$$

Since θ terminates in QI, x must equal 5.

$$\cos \theta = \frac{x}{r} = \frac{5}{13}$$

$$\tan \theta = \frac{y}{x} = \frac{12}{5}$$

$$\cot \theta = \frac{x}{y} = \frac{5}{12}$$

$$\sec \theta = \frac{r}{x} = \frac{13}{5}$$

$$\csc \theta = \frac{r}{y} = \frac{13}{12}$$

45. θ terminates in QIV. Therefore, x is positive and y is negative.

$$\cos \theta = \frac{x}{r} = \frac{24}{25}$$

Therefore, $x = 24$ and $r = 25$.

$$x^2 + y^2 = r^2$$

$$(24)^2 + y^2 = (25)^2$$

$$576 + y^2 = 625$$

$$y^2 = 49$$

$$y = \pm 7$$

Therefore, $y = -7$

This problem continued on the next page.

$$\sin \theta = \frac{y}{r} = -\frac{7}{25}$$

$$\tan \theta = \frac{y}{x} = -\frac{7}{24}$$

$$\cot \theta = \frac{x}{y} = -\frac{24}{7}$$

$$\sec \theta = \frac{r}{x} = \frac{25}{24}$$

$$\csc \theta = \frac{r}{y} = -\frac{25}{7}$$

49. θ terminates in QIII. Therefore, both x and y are negative.

$$\sin \theta = \frac{y}{r} = \frac{-20}{29}$$

Therefore, $y = -20$ and $r = 29$

$$x^2 + y^2 = r^2$$

$$x^2 + (-20)^2 = (29)^2$$

$$x^2 + 400 = 841$$

$$x^2 = 441$$

$$x = \pm 21$$

Therefore, $x = -21$.

$$\cos \theta = \frac{x}{r} = -\frac{21}{29}$$

$$\tan \theta = \frac{y}{x} = \frac{-20}{-21} = \frac{20}{21}$$

$$\cot \theta = \frac{x}{y} = \frac{-21}{-20} = \frac{21}{20}$$

$$\sec \theta = \frac{r}{x} = -\frac{29}{21}$$

$$\csc \theta = \frac{r}{y} = -\frac{29}{20}$$

53. $\cot \theta$ is positive in QI and QIII.

$\cos \theta$ is positive in QI and QIV.

Therefore, θ is in QI and both x and y are positive.

$$\cot \theta = \frac{x}{y} = \frac{1}{2}$$

Therefore, $x = 1$ and $y = 2$.

$$x^2 + y^2 = r^2$$
$$1^2 + 2^2 = r^2$$
$$1 + 4 = r^2$$
$$r^2 = 5$$
$$r = \sqrt{5} \text{ because } r \text{ is always positive}$$

$$\sin\theta = \frac{y}{r} = \frac{2}{\sqrt{5}}$$

$$\cos\theta = \frac{x}{r} = \frac{1}{\sqrt{5}}$$

$$\tan\theta = \frac{y}{x} = \frac{2}{1} = 2$$

$$\sec\theta = \frac{r}{x} = \frac{\sqrt{5}}{1} = \sqrt{5}$$

$$\csc\theta = \frac{r}{y} = \frac{\sqrt{5}}{2}$$

57. In QII, x is negative and y is positive.

$$\cot\theta = \frac{x}{y} = \frac{-2}{1}.$$

Therefore, $x = -2$ and $y = 1$.

From Problem 53, we found that $r = \sqrt{5}$.

$$\sin\theta = \frac{y}{r} = \frac{1}{\sqrt{5}}$$

$$\cos\theta = \frac{x}{r} = -\frac{2}{\sqrt{5}}$$

$$\tan\theta = \frac{y}{x} = -\frac{1}{2}$$

$$\sec\theta = \frac{r}{x} = -\frac{\sqrt{5}}{2}$$

$$\csc\theta = \frac{r}{y} = \frac{\sqrt{5}}{1} = \sqrt{5}$$

61. $\sin\theta = \dfrac{y}{r} = \dfrac{1}{1}$

$x^2 + y^2 = r^2$

$x^2 + 1^2 = 1^2$

$x^2 + 1 = 1$

$x^2 = 0$

$x = 0$

Therefore, the point $(0, 1)$ lies on the terminal side of θ and θ must be $90°$.

65. The terminal side of θ lies in QI. Therefore, the point $(1, 2)$ lies along $y = 2x$ in QI.

$(x, y) = (1, 2)$

$x = 1$ and $y = 2$

$r^2 = x^2 + y^2$ $\qquad\qquad$ $\sin\theta = \dfrac{y}{r} = \dfrac{2}{\sqrt{5}}$

$\quad = 1^2 + 2^2$

$\quad = 1 + 4$ $\qquad\qquad\qquad$ $\cos\theta = \dfrac{x}{r} = \dfrac{1}{\sqrt{5}}$

$\quad = 5$

$r = \sqrt{5}$

69. The point $(1, 1)$ lies on the terminal side of $45°$ in QI. Therefore, $x = 1$ and $y = 1$.

$r^2 = x^2 + y^2$

$\quad = 1^2 + 1^2$

$\quad = 2$

$r = \sqrt{2}$

$\cos 45° = \dfrac{x}{r} = \dfrac{1}{\sqrt{2}}$

The point $(1, -1)$ lies on the terminal side of $-45°$ in QIV. Therefore, $x = 1$ and $y = -1$.

$r^2 = 1^2 + (-1)^2$

$\quad = 1 + 1$

$\quad = 2$

$r = \sqrt{2}$ and $\cos(-45°) = \dfrac{x}{r} = \dfrac{1}{\sqrt{2}}$

Problem Set 1.4

1. $\dfrac{1}{7}$

5. $\dfrac{1}{-1/\sqrt{2}} = 1 \cdot \dfrac{\sqrt{2}}{-1} = -\sqrt{2}$

9. $\csc\theta = \dfrac{1}{\sin\theta}$

$= \dfrac{1}{4/5}$

$= \dfrac{5}{4}$

13. $\cot\theta = \dfrac{1}{\tan\theta}$

$= \dfrac{1}{a}, \ a \neq 0$

17. $\cot\theta = \dfrac{\cos\theta}{\sin\theta}$

$= \dfrac{-12/13}{-5/13}$

$= \dfrac{12}{5}$

21. $\tan^3\theta = (\tan\theta)^3$

$= (2)^3$

$= 8$

25. $\sec\theta = \dfrac{1}{\cos\theta}$

$= \dfrac{1}{-5/13}$

$= -\dfrac{13}{5}$

29. $\cos\theta = \pm\sqrt{1 - \sin^2\theta}$

$$= \pm\sqrt{1 - \left(\frac{1}{3}\right)^2}$$

$$= \pm\sqrt{1 - \frac{1}{9}}$$

$$= \pm\sqrt{\frac{8}{9}}$$

$$= \pm\frac{2\sqrt{2}}{3}$$

Since θ terminates in QII, $\cos\theta$ is negative. Therefore, $\cos\theta = -\dfrac{2\sqrt{2}}{3}$.

33. $\sin\theta = \pm\sqrt{1 - \cos^2\theta}$

$$= \pm\sqrt{1 - \left(\frac{\sqrt{3}}{2}\right)^2}$$

$$= \pm\sqrt{1 - \frac{3}{4}}$$

$$= \pm\sqrt{\frac{1}{4}}$$

$$= \pm\frac{1}{2}$$

Since θ terminates in QI, $\sin\theta$ is positive.

Therefore, $\sin\theta = \dfrac{1}{2}$

37. $\cos\theta = \pm\sqrt{1 - \sin^2\theta}$

$$= \pm\sqrt{1 - \left(\frac{1}{3}\right)^2}$$

$$= \pm\sqrt{1 - \frac{1}{9}}$$

$$= \pm\sqrt{\frac{8}{9}}$$

$$= \pm\frac{2\sqrt{2}}{3}$$

Since θ terminates in QI, $\cos\theta$ is positive.

Therefore, $\cos\theta = \dfrac{2\sqrt{2}}{3}$.

$$\tan\theta = \dfrac{\sin\theta}{\cos\theta}$$

$$= \dfrac{1/3}{2\sqrt{2}/3}$$

$$= \dfrac{1}{2\sqrt{2}}$$

41. $\csc\theta = \pm\sqrt{1+\cot^2\theta}$ Pythagorean identity

$$= \pm\sqrt{1+\left(-\dfrac{21}{20}\right)^2}$$ Substitute given value

$$= \pm\sqrt{1+\dfrac{441}{400}}$$ Simplify

$$= \pm\sqrt{\dfrac{841}{400}}$$

$$= \pm\dfrac{29}{20}$$

Since $\cot\theta$ is negative and $\sin\theta$ is positive, θ must lie in QII.

Therefore, $\csc\theta = \dfrac{29}{20}$.

45. In QIII and QIV, $\sin\theta$ is negative, but θ is not in QIII. Therefore, θ must be in QIV.

$$\csc\theta = \dfrac{1}{\sin\theta}$$

$$= \dfrac{1}{-1/2}$$

$$= -2$$

$$\cos\theta = \pm\sqrt{1-\sin^2\theta}$$

$$= \pm\sqrt{1-\left(-\dfrac{1}{2}\right)^2}$$

$$= \pm\sqrt{1-\dfrac{1}{4}}$$

$$= \pm\sqrt{\dfrac{3}{4}}$$

$$= \pm\dfrac{\sqrt{3}}{2}$$

This problem continued on the next page.

$$\cos \theta = \frac{\sqrt{3}}{2} \text{ because } \cos \theta \text{ is positive in QIV}$$

$$\tan \theta = \frac{\sin \theta}{\cos \theta}$$

$$= \frac{-1/2}{\sqrt{3}/2}$$

$$= -\frac{1}{\sqrt{3}}$$

$$\cot \theta = \frac{1}{\tan \theta}$$

$$= \frac{1}{-1\sqrt{3}}$$

$$= -\sqrt{3}$$

$$\sec \theta = \frac{1}{\cos \theta}$$

$$= \frac{1}{\sqrt{3}/2}$$

$$= \frac{2}{\sqrt{3}}$$

All six ratios are:

$$\sin \theta = -\frac{1}{2} \qquad \cot \theta = -\sqrt{3}$$

$$\cos \theta = \frac{\sqrt{3}}{2} \qquad \sec \theta = \frac{2}{\sqrt{3}}$$

$$\tan \theta = -\frac{1}{\sqrt{3}} \qquad \csc \theta = -2$$

49. $$\sin \theta = \pm \sqrt{1 - \cos^2 \theta}$$

$$= \pm \sqrt{1 - \left(\frac{2}{\sqrt{13}}\right)^2}$$

$$= \pm \sqrt{1 - \frac{4}{13}}$$

$$= \pm \sqrt{\frac{9}{13}}$$

$$= \pm \frac{3}{\sqrt{13}}$$

$$\sin \theta = -\frac{3}{\sqrt{13}} \quad \text{because } \sin \theta \text{ is negative in QIV}$$

$$\tan \theta = \frac{\sin \theta}{\cos \theta} \qquad\qquad \cot \theta = \frac{1}{\tan \theta}$$

$$= \frac{-3/\sqrt{13}}{2/\sqrt{13}} \qquad\qquad = \frac{1}{-3/2}$$

$$= -\frac{3}{2} \qquad\qquad\qquad = -\frac{2}{3}$$

$$\sec \theta = \frac{1}{\cos \theta} \qquad\qquad \csc \theta = \frac{1}{\sin \theta}$$

$$= \frac{1}{2/\sqrt{13}} \qquad\qquad = \frac{1}{-3/\sqrt{13}}$$

$$= \frac{\sqrt{13}}{2} \qquad\qquad\qquad = -\frac{\sqrt{13}}{3}$$

All six ratios are:

$$\sin \theta = -\frac{3}{\sqrt{13}} \qquad\qquad \cot \theta = -\frac{2}{3}$$

$$\cos \theta = \frac{2}{\sqrt{13}} \qquad\qquad \sec \theta = \frac{\sqrt{13}}{2}$$

$$\tan \theta = -\frac{3}{2} \qquad\qquad \csc \theta = -\frac{\sqrt{13}}{3}$$

53. $\sin \theta = \dfrac{1}{\csc \theta}$

$$\sin \theta = \frac{1}{a}, \ a \neq 0$$

$$\cos \theta = \pm \sqrt{1 - \sin^2 \theta}$$

$$= \pm \sqrt{1 - \left(\frac{1}{a}\right)^2}$$

$$= \pm \sqrt{1 - \frac{1}{a^2}}$$

$$= \pm \sqrt{\frac{a^2 - 1}{a^2}}$$

$$= \pm \frac{\sqrt{a^2 - 1}}{a}$$

$$= \frac{\sqrt{a^2 - 1}}{a} \quad \text{because } \theta \text{ is in QI}$$

This problem continued on the next page.

$$\tan\theta = \frac{\sin\theta}{\cos\theta} \qquad\qquad \cot\theta = \frac{1}{\tan\theta}$$

$$= \frac{1/a}{\sqrt{a^2-1}/a} \qquad\qquad = \frac{1}{1/\sqrt{a^2-1}}$$

$$= \frac{1}{\sqrt{a^2-1}} \qquad\qquad = \sqrt{a^2-1}$$

$$\sec\theta = \frac{1}{\cos\theta} = \frac{1}{\sqrt{a^2-1}/a} = \frac{a}{\sqrt{a^2-1}}$$

All six ratios are:

$$\sin\theta = \frac{1}{a} \qquad\qquad \cot\theta = \sqrt{a^2-1}$$

$$\cos\theta = \frac{\sqrt{a^2-1}}{a} \qquad\qquad \sec\theta = \frac{a}{\sqrt{a^2-1}}$$

$$\tan\theta = \frac{1}{\sqrt{a^2-1}} \qquad\qquad \csc\theta = a$$

57. $\cos\theta = \dfrac{1}{\sec\theta}$

$$= \frac{1}{-1.24}$$

$$= -0.806$$

$$= -0.81 \text{ (rounded to the nearest hundredth)}$$

$$\sin\theta = \pm\sqrt{1 - \cos^2\theta}$$

$$= \pm\sqrt{1 - (-0.81)^2}$$

$$= \pm\sqrt{1 - 0.6561}$$

$$= \pm\sqrt{0.3439}$$

$$= \pm 0.586$$

$$\sin\theta = 0.59 \text{ because } \sin\theta \text{ is positive in QII}$$

$$\tan \theta = \frac{\sin \theta}{\cos \theta} \qquad\qquad \cot \theta = \frac{1}{\tan \theta}$$

$$= \frac{0.59}{-0.81} \qquad\qquad = \frac{1}{-0.73}$$

$$= -0.728 \qquad\qquad = -1.369$$

$$= -0.73 \qquad\qquad = -1.37$$

$$\csc \theta = \frac{1}{\sin \theta}$$

$$= \frac{1}{0.59}$$

$$= 1.694$$

$$= 1.69$$

All six ratios are:

$$\sin \theta = 0.59 \qquad\qquad \cot \theta = -1.37$$

$$\cos \theta = -0.81 \qquad\qquad \sec \theta = -1.24$$

$$\tan \theta = -0.73 \qquad\qquad \csc \theta = 1.69$$

61. This line passes through (0, 0) and (1, m).

$$\text{slope} = \frac{y_2 - y_1}{x_2 - x_1}$$

$$= \frac{m - 0}{1 - 0}$$

$$\text{slope} = \frac{m}{1} \text{ or } m$$

Problem Set 1.5

1. $\cos \theta = \pm \sqrt{1 - \sin^2 \theta}$ Pythagorean identity

5. $\sec \theta = \dfrac{1}{\cos \theta}$ Reciprocal identity

9. $\csc \theta \cot \theta = \dfrac{1}{\sin \theta} \cdot \dfrac{\cos \theta}{\sin \theta}$ Reciprocal and ratio identities

$$= \frac{\cos \theta}{\sin^2 \theta} \qquad\qquad \text{Multiplication}$$

13. $\dfrac{\sec\theta}{\csc\theta} = \dfrac{1/\cos\theta}{1/\sin\theta}$ Reciprocal identities

$\qquad\qquad = \dfrac{\sin\theta}{\cos\theta}$ Division of fractions

17. $\dfrac{\tan\theta}{\cot\theta} = \dfrac{\sin\theta/\cos\theta}{\cos\theta/\sin\theta}$ Ratio identities

$\qquad\qquad = \dfrac{\sin^2\theta}{\cos^2\theta}$ Division of fractions

21. $\tan\theta + \sec\theta = \dfrac{\sin\theta}{\cos\theta} + \dfrac{1}{\cos\theta}$ Ratio and reciprocal identities

$\qquad\qquad\qquad = \dfrac{\sin\theta + 1}{\cos\theta}$ Addition of fractions

25. $\sec\theta - \tan\theta\,\sin\theta = \dfrac{1}{\cos\theta} - \dfrac{\sin\theta}{\cos\theta}\cdot\sin\theta$ Reciprocal and ratio identities

$\qquad\qquad\qquad\qquad = \dfrac{1}{\cos\theta} - \dfrac{\sin^2\theta}{\cos\theta}$ Multiplication of fractions

$\qquad\qquad\qquad\qquad = \dfrac{1 - \sin^2\theta}{\cos\theta}$ Subtraction of fractions

$\qquad\qquad\qquad\qquad = \dfrac{\cos^2\theta}{\cos\theta}$ Pythagorean identity

$\qquad\qquad\qquad\qquad = \cos\theta$ Division

29. $\dfrac{1}{\sin\theta} - \dfrac{1}{\cos\theta} = \dfrac{1}{\sin\theta}\cdot\dfrac{\cos\theta}{\cos\theta} - \dfrac{1}{\cos\theta}\cdot\dfrac{\sin\theta}{\sin\theta}$ LCD is $\sin\theta\cos\theta$

$\qquad\qquad\qquad\quad = \dfrac{\cos\theta}{\sin\theta\cos\theta} - \dfrac{\sin\theta}{\sin\theta\cos\theta}$ Multiplication

$\qquad\qquad\qquad\quad = \dfrac{\cos\theta - \sin\theta}{\sin\theta\cos\theta}$ Subtraction of fractions

33. $\dfrac{1}{\sin\theta} - \sin\theta = \dfrac{1}{\sin\theta} - \sin\theta\cdot\dfrac{\sin\theta}{\sin\theta}$ LCD is $\sin\theta$

$\qquad\qquad\qquad = \dfrac{1}{\sin\theta} - \dfrac{\sin^2\theta}{\sin\theta}$ Multiplication

$$= \frac{1 - \sin^2\theta}{\sin\theta} \qquad \text{Subtraction of fractions}$$

$$= \frac{\cos^2\theta}{\sin\theta} \qquad \text{Pythagorean identity}$$

37. $(2\cos\theta + 3)(4\cos\theta - 5) = 8\cos^2\theta - 10\cos\theta + 12\cos\theta - 15$

$$= 8\cos^2\theta + 2\cos\theta - 15$$

41. $(1 - \tan\theta)(1 + \tan\theta) = 1 + \tan\theta - \tan\theta - \tan^2\theta$

$$= 1 - \tan^2\theta$$

45. $(\sin\theta - 4)^2 = (\sin\theta - 4)(\sin\theta - 4)$

$$= \sin^2\theta - 4\sin\theta - 4\sin\theta + 16$$

$$= \sin^2\theta - 8\sin\theta + 16$$

49. $\sin\theta\sec\theta\cot\theta = \sin\theta \cdot \dfrac{1}{\cos\theta} \cdot \dfrac{\cos\theta}{\sin\theta}$ \qquad Reciprocal and ratio identities

$$= \frac{\sin\theta\cos\theta}{\cos\theta\sin\theta} \qquad \text{Multiplication}$$

$$= 1 \qquad \text{Division of common factors}$$

53. $\dfrac{\csc\theta}{\cot\theta} = \dfrac{1/\sin\theta}{\cos\theta/\sin\theta}$ \qquad Reciprocal and ratio identities

$$= \frac{\sin\theta}{\sin\theta\cos\theta} \qquad \text{Division of fractions}$$

$$= \frac{1}{\cos\theta} \qquad \text{Division of common factor}$$

$$= \sec\theta \qquad \text{Reciprocal identity}$$

57. $\dfrac{\sec\theta\cot\theta}{\csc\theta} = \dfrac{\frac{1}{\cos\theta} \cdot \frac{\cos\theta}{\sin\theta}}{\frac{1}{\sin\theta}}$ \qquad Reciprocal and ratio identities

$$= \frac{\cos\theta\sin\theta}{\cos\theta\sin\theta} \qquad \text{Division of fractions}$$

$$= 1 \qquad \text{Division of common factors}$$

61. $\tan\theta + \cot\theta = \dfrac{\sin\theta}{\cos\theta} + \dfrac{\cos\theta}{\sin\theta}$ Ratio and reciprocal identities

$\qquad\qquad = \dfrac{\sin\theta}{\cos\theta} \cdot \dfrac{\boldsymbol{\sin\theta}}{\boldsymbol{\sin\theta}} + \dfrac{\cos\theta}{\sin\theta} \cdot \dfrac{\boldsymbol{\cos\theta}}{\boldsymbol{\cos\theta}}$ LCD is $\sin\theta\cos\theta$

$\qquad\qquad = \dfrac{\sin^2\theta}{\sin\theta\cos\theta} + \dfrac{\cos^2\theta}{\sin\theta\cos\theta}$ Multiplication

$\qquad\qquad = \dfrac{\sin^2\theta + \cos^2\theta}{\sin\theta\cos\theta}$ Addition of fractions

$\qquad\qquad = \dfrac{1}{\sin\theta\cos\theta}$ Pythagorean identity

$\qquad\qquad = \dfrac{1}{\sin\theta} \cdot \dfrac{1}{\cos\theta}$ Property of fractions

$\qquad\qquad = \csc\theta\sec\theta$ Reciprocal identities

65. $\csc\theta\tan\theta - \cos\theta = \dfrac{1}{\sin\theta} \cdot \dfrac{\sin\theta}{\cos\theta} - \cos\theta$ Reciprocal and ratio identities

$\qquad\qquad = \dfrac{\sin\theta}{\sin\theta\cos\theta} - \cos\theta$ Multiplication

$\qquad\qquad = \dfrac{1}{\cos\theta} - \cos\theta$ Division of common factor

$\qquad\qquad = \dfrac{1}{\cos\theta} - \cos\theta \cdot \dfrac{\boldsymbol{\cos\theta}}{\boldsymbol{\cos\theta}}$ LCD is $\cos\theta$

$\qquad\qquad = \dfrac{1}{\cos\theta} - \dfrac{\cos^2\theta}{\cos\theta}$ Multiplication

$\qquad\qquad = \dfrac{1 - \cos^2\theta}{\cos\theta}$ Subtraction of fractions

$\qquad\qquad = \dfrac{\sin^2\theta}{\cos\theta}$ Pythagorean identity

69. $(\sin\theta + 1)(\sin\theta - 1) = \sin^2\theta - 1$ Multiplication

$\qquad\qquad = 1 - \cos^2\theta - 1$ Pythagorean identity

$\qquad\qquad = -\cos^2\theta$ Addition

73. $(\sin\theta - \cos\theta)^2 - 1 = \sin^2\theta - 2\sin\theta\cos\theta + \cos^2\theta - 1$ Multiplication

$\qquad\qquad = -2\sin\theta\cos\theta + (\sin^2\theta + \cos^2\theta) - 1$ Commutative property

$\qquad\qquad = -2\sin\theta\cos\theta + 1 - 1$ Pythagorean identity

$\qquad\qquad = -2\sin\theta\cos\theta$ Addition

77. $\sin\theta\,(\sec\theta + \cot\theta) = \sin\theta\,\sec\theta + \sin\theta\,\cot\theta$ Multiplication

$$= \sin\theta \cdot \frac{1}{\cos\theta} + \sin\theta \cdot \frac{\cos\theta}{\sin\theta} \qquad \text{Reciprocal and ratio identities}$$

$$= \frac{\sin\theta}{\cos\theta} + \frac{\sin\theta\,\cos\theta}{\sin\theta} \qquad \text{Multiplication}$$

$$= \tan\theta + \cos\theta \qquad \text{Ratio identity and division}$$
$$\text{of common factor}$$

81. $\sqrt{x^2 + 4} = \sqrt{(2\tan\theta)^2 + 4}$ Substitute given value

$$= \sqrt{4\tan^2\theta + 4} \qquad \text{Multiply}$$

$$= \sqrt{4(\tan^2\theta + 1)} \qquad \text{Factor}$$

$$= \sqrt{4\sec^2\theta} \qquad \text{Pythagorean identity}$$

$$= |2\sec\theta| = 2|\sec\theta| \qquad \text{Simplify}$$

85. $\sqrt{4x^2 + 16} = \sqrt{4\,(2\tan\theta)^2 + 16}$ Substitute given value

$$= \sqrt{16\tan^2\theta + 16} \qquad \text{Multiply}$$

$$= \sqrt{16\,(\tan^2\theta + 1)} \qquad \text{Factor}$$

$$= \sqrt{16\sec^2\theta} \qquad \text{Pythagorean identity}$$

$$= |4\sec\theta| = 4|\sec\theta| \qquad \text{Simplify}$$

Chapter 1 Test

1. $x^2 + (3)^2 = (6)^2$ Pythagorean theorem

   $x^2 + 9 = 36$ Simplify

   $x^2 = 27$ Subtract 9 from both sides

   $x = \pm 3\sqrt{3}$ Take square root of both sides

   $x = 3\sqrt{3}$ because x must be positive

2. $x^2 + 4^2 = (x + 2)^2$ Pythagorean theorem

   $x^2 + 16 = x^2 + 4x + 4$ Simplify

   $12 = 4x$ Subtract 4 and x^2 from both sides

   $3 = x$ Divide both sides by 4

3. $\tan 45° = \dfrac{h}{s}$ Tangent relationship

   $1 = \dfrac{h}{5\sqrt{3}}$ Substitute known values

   $h = 5\sqrt{3}$ Multiply both sides by $5\sqrt{3}$

   $\cos 45° = \dfrac{s}{r}$ Cosine relationship

   $\dfrac{\sqrt{2}}{2} = \dfrac{5\sqrt{3}}{r}$ Substitute known values

   $\dfrac{\sqrt{2}}{2} r = 5\sqrt{3}$ Solve for r

   $r = \dfrac{10\sqrt{3}}{\sqrt{2}}$

   $= 5\sqrt{6}$ Rationalize the denominator

   $\tan 60° = \dfrac{h}{y}$ Tangent relationship

   $\sqrt{3} = \dfrac{5\sqrt{3}}{y}$ Substitute known values

   $\sqrt{3}\, y = 5\sqrt{3}$ Solve for y

   $y = 5$

$$\sin 30° = \frac{y}{x}$$ Sine relationship

$$\frac{1}{2} = \frac{5}{x}$$ Substitute known values

$$\frac{1}{2}x = 5$$ Solve for x

$$x = 10$$

4. $\sin 30° = \dfrac{y}{x}$ Sine relationship

$$\frac{1}{2} = \frac{3}{x}$$ Substitute known values

$$x = 6$$ Solve for x

$$\cos 30° = \frac{h}{x}$$ Cosine relationship

$$\frac{\sqrt{3}}{2} = \frac{h}{6}$$ Substitute known values

$$h = 3\sqrt{3}$$ Solve for h

$$\tan 45° = \frac{s}{h}$$ Tangent relationship

$$1 = \frac{s}{3\sqrt{3}}$$ Substitute known values

$$s = 3\sqrt{3}$$ Solve for s

$$\sin 45° = \frac{h}{r}$$ Sine relationship

$$\frac{\sqrt{2}}{2} = \frac{3\sqrt{3}}{r}$$ Substitute known values

$$\frac{\sqrt{2}}{2}r = 3\sqrt{3}$$ Solve for r

$$r = \frac{6\sqrt{3}}{\sqrt{2}}$$

$$= 3\sqrt{6}$$ Rationalize the denominator

5. $(AB)^2 + (BC)^2 = (AC)^2$ Pythagorean theorem

$(AB)^2 + (3)^2 = (5)^2$ Substitute given values

$(AB)^2 + 9 = 25$ Simplify

$(AB)^2 = 16$ Subtract 9 from both sides

This problem continued on the next page.

$$AB = \pm 4 \qquad \text{Take square root of both sides}$$

$$AB = 4 \quad \text{because it must be positive}$$

$$(DB)^2 = (DA)^2 + (AB)^2 \qquad \text{Pythagorean theorem}$$

$$(DB)^2 = (6)^2 + (4)^2 \qquad \text{Substitute given values}$$

$$(DB)^2 = 36 + 16 \qquad \text{Simplify}$$

$$(DB)^2 = 52$$

$$DB = \pm \sqrt{52} \qquad \text{Take square root of both sides}$$

$$DB = \sqrt{52} \quad \text{or} \quad 7.21 \quad \text{because it must be positive}$$

6. One complete revolution takes 12 hours or 360°. Let θ represent the number of degrees a clock moves in 3 hours, then:

$$\frac{\theta}{360°} = \frac{3 \text{ hr}}{12 \text{ hr}}$$

$$\frac{\theta}{360°} = \frac{1}{4}$$

$$\theta = 90°$$

7. The shortest side is $\frac{1}{2}(5)$ or $\frac{5}{2}$.

The medium side is $\frac{5}{2}(\sqrt{3})$ or $\frac{5\sqrt{3}}{2}$.

8. The x-intercept is $(3, 0)$ and the y-intercept is $(0, 2)$.

9.
$$r = \sqrt{(x_2 - x_1)^2 + (y_2 - y_1)^2} \qquad \text{Distance formula}$$

$$= \sqrt{(4 + 1)^2 + (-2 - 10)^2} \qquad \text{Substitute given values}$$

$$= \sqrt{(5)^2 + (-12)^2} \qquad \text{Simplify}$$

$$= \sqrt{25 + 144}$$

$$= \sqrt{169}$$

$$= 13$$

10.
$$d = \sqrt{(a - 0)^2 + (b - 0)^2} \qquad \text{Distance formula}$$

$$= \sqrt{a^2 + b^2} \qquad \text{Simplify}$$

11. $\sqrt{(x+2)^2 + (1-3)^2} = \sqrt{13}$ Distance formula

$\sqrt{(x+2)^2 + 4} = \sqrt{13}$ Simplify

$(x+2)^2 + 4 = 13$ Square both sides

$(x+2)^2 = 9$ Subtract 4 from both sides

$x + 2 = \pm 3$ Take square root of both sides

$x = -2 \pm 3$ Solve for x

$x = -5 \text{ or } x = 1$

12. A point on the terminal side of $90°$ is $(0, 1)$ and $r = 1$.

$$\sin 90° = \frac{y}{r} = \frac{1}{1} = 1$$

$$\cos 90° = \frac{x}{r} = \frac{0}{1} = 0$$

$$\tan 90° = \frac{y}{x} = \frac{1}{0} \text{ is undefined}$$

13. A point on the terminal side of $-45°$ is $(1, -1)$ and $r = \sqrt{2}$.

$$\sin 45° = \frac{y}{r} = \frac{-1}{\sqrt{2}} = -\frac{1}{\sqrt{2}}$$

$$\cos 45° = \frac{x}{r} = \frac{1}{\sqrt{2}}$$

$$\tan 45° = \frac{y}{x} = \frac{-1}{1} = -1$$

14. $\sin \theta$ is negative in QIII and QIV.

$\cos \theta$ is positive in QI and QIV.

Therefore, θ must lie in QIV.

15. $\csc \theta$ is positive in QI and QII.

$\cos \theta$ is negative in QII and QIII.

Therefore, θ must lie in QII.

16. $(x, y) = (-6, 8)$

$x = -6$ and $y = 8$

$$r = \sqrt{x^2 + y^2}$$
$$= \sqrt{(-6)^2 + 8^2}$$
$$= \sqrt{36 + 64}$$
$$= \sqrt{100}$$
$$= 10$$

$$\sin \theta = \frac{y}{r} = \frac{8}{10} = \frac{4}{5} \qquad\qquad \cot \theta = \frac{x}{y} = \frac{-6}{8} = -\frac{3}{4}$$

$$\cos \theta = \frac{x}{r} = \frac{-6}{10} = -\frac{3}{5} \qquad\qquad \sec \theta = \frac{r}{x} = \frac{10}{-6} = -\frac{5}{3}$$

$$\tan \theta = \frac{y}{x} = \frac{8}{-6} = -\frac{4}{3} \qquad\qquad \csc \theta = \frac{r}{y} = \frac{10}{8} = \frac{5}{4}$$

17. $(x, y) = (-3, -1)$

$x = -3$ and $y = -1$

$$r = \sqrt{x^2 + y^2}$$
$$= \sqrt{(-3)^2 + (-1)^2}$$
$$= \sqrt{9 + 1}$$
$$= \sqrt{10}$$

$$\sin \theta = \frac{y}{r} = \frac{-1}{\sqrt{10}} = -\frac{1}{\sqrt{10}} \qquad\qquad \cot \theta = \frac{x}{y} = \frac{-3}{-1} = 3$$

$$\cos \theta = \frac{x}{r} = \frac{-3}{\sqrt{10}} = -\frac{3}{\sqrt{10}} \qquad\qquad \sec \theta = \frac{r}{x} = \frac{\sqrt{10}}{-3} = -\frac{\sqrt{10}}{3}$$

$$\tan \theta = \frac{y}{x} = \frac{-1}{-3} = \frac{1}{3} \qquad\qquad \csc \theta = \frac{r}{y} = \frac{\sqrt{10}}{-1} = -\sqrt{10}$$

18. If $\sin \theta = \dfrac{1}{2}$ and θ terminates in QII, then:

$$\cos \theta = -\sqrt{1 - \left(\frac{1}{2}\right)^2}$$
$$= -\sqrt{\frac{3}{4}}$$
$$= -\frac{\sqrt{3}}{2}$$

$$\tan\theta = \frac{\sin\theta}{\cos\theta} = \frac{1/2}{-\sqrt{3}/2} = -\frac{1}{\sqrt{3}}$$

$$\cot\theta = \frac{1}{\tan\theta} = \frac{1}{-1/\sqrt{3}} = \sqrt{3}$$

$$\sec\theta = \frac{1}{\cos\theta} = \frac{1}{-\sqrt{3}/2} = -\frac{2}{\sqrt{3}}$$

$$\csc\theta = \frac{1}{\sin\theta} = \frac{1}{1/2} = 2$$

19. If $\tan\theta = \dfrac{12}{5}$ and θ terminates in QIII, then:

$$\sec^2\theta = \tan^2\theta + 1$$

$$= \frac{144}{25} + 1$$

$$= \frac{169}{25}$$

$$\sec\theta = -\frac{13}{5} \text{ because } \sec\theta \text{ is negative in QIII}$$

$$\cot\theta = \frac{1}{\tan\theta} = \frac{1}{12/5} = \frac{5}{12}$$

$$\cos\theta = \frac{1}{\sec\theta} = \frac{1}{-13/5} = -\frac{5}{13}$$

$$\frac{\sin\theta}{\cos\theta} = \tan\theta \qquad\qquad \csc\theta = \frac{1}{\sin\theta}$$

$$\frac{\sin\theta}{-5/13} = \frac{12}{5} \qquad\qquad = \frac{1}{-12/13}$$

$$\sin\theta = -\frac{12}{13} \qquad\qquad = -\frac{13}{12}$$

20. A point on the terminal side of θ in QIV is $(1, -2)$.

$(x, y) = (1, -2)$

$x = 1$ and $y = -2$

$$r = \sqrt{x^2 + y^2}$$

$$= \sqrt{1^2 + (-2)^2}$$

$$= \sqrt{1 + 4}$$

$$= \sqrt{5}$$

$$\sin\theta = \frac{y}{r} = \frac{-2}{\sqrt{5}} \qquad\qquad \cos\theta = \frac{x}{r} = \frac{1}{\sqrt{5}}$$

21.
$$\csc \theta = \frac{1}{\sin \theta} \qquad \text{Reciprocal identity}$$
$$= \frac{1}{-3/4} \qquad \text{Substitute given value}$$
$$= -\frac{4}{3} \qquad \text{Simplify}$$

22.
$$\cos \theta = \frac{1}{\sec \theta} \qquad \text{Reciprocal identity}$$
$$= \frac{1}{-2}$$
$$= -\frac{1}{2}$$

23. $\sin^3 \theta = \left(\frac{1}{3}\right)^3 = \frac{1}{27}$

24.
$$\cos \theta = \frac{1}{\sec \theta} \qquad \text{Reciprocal identity}$$
$$= \frac{1}{3}$$

$$\sin \theta = -\sqrt{1 - \cos^2 \theta} \text{ because } \theta \text{ is in QIV}$$
$$= -\sqrt{1 - \frac{1}{9}}$$
$$= -\sqrt{\frac{8}{9}}$$
$$= -\frac{2\sqrt{2}}{3}$$

$$\tan \theta = \frac{\sin \theta}{\cos \theta} \qquad \text{Ratio identity}$$
$$= -\frac{2\sqrt{2}/3}{1/3}$$
$$= -2\sqrt{2}$$

25.
$$\cos\theta = \pm\sqrt{1 - \sin^2\theta} \qquad \text{Pythagorean identity}$$

$$= \pm\sqrt{1 - \left(\frac{1}{a}\right)^2} \qquad \text{Substitute given value}$$

$$= \pm\sqrt{1 - \frac{1}{a^2}} \qquad \text{Simplify}$$

$$= \pm\sqrt{\frac{a^2 - 1}{a^2}}$$

$$= \pm\frac{\sqrt{a^2 - 1}}{a}$$

$$\cos\theta = \frac{\sqrt{a^2 - 1}}{a} \quad \text{because } \cos\theta \text{ is positive in QI}$$

$$\csc\theta = \frac{1}{\sin\theta} \qquad \text{Reciprocal identity}$$

$$= \frac{1}{1/a} \qquad \text{Substitute given value}$$

$$= a \qquad \text{Simplify}$$

$$\cot\theta = \frac{\cos\theta}{\sin\theta} \qquad \text{Ratio identity}$$

$$= \frac{\sqrt{a^2 - 1}/a}{1/a} \qquad \text{Substitute given values}$$

$$= \sqrt{a^2 - 1} \qquad \text{Simplify}$$

26.
$$(\sin\theta + 3)(\sin\theta - 7) = \sin^2\theta - 7\sin\theta + 3\sin\theta - 21$$
$$= \sin^2\theta - 4\sin\theta - 21$$

27.
$$(\cos\theta - \sin\theta)^2 = \cos^2\theta - 2\sin\theta\cos\theta + \sin^2\theta \qquad \text{Multiplication}$$
$$= (\cos^2\theta + \sin^2\theta) - 2\sin\theta\cos\theta \qquad \text{Commutative Property}$$
$$= 1 - 2\sin\theta\cos\theta \qquad \text{Pythagorean identity}$$

28.
$$\frac{1}{\sin\theta} - \sin\theta = \frac{1}{\sin\theta} - \frac{\sin^2\theta}{\sin\theta} \qquad \text{LCD is } \sin\theta$$

$$= \frac{1 - \sin^2\theta}{\sin\theta} \qquad \text{Subtraction of fractions}$$

$$= \frac{\cos^2\theta}{\sin\theta} \qquad \text{Pythagorean identity}$$

29. $\dfrac{\cot \theta}{\csc \theta} = \dfrac{\cos \theta / \sin \theta}{1/\sin \theta}$ Ratio and reciprocal identities

$\qquad = \dfrac{\cos \theta \sin \theta}{\sin \theta}$ Division of fractions

$\qquad = \cos \theta$ Division of common factor

30. $\cot \theta + \tan \theta = \dfrac{\cos \theta}{\sin \theta} + \dfrac{\sin \theta}{\cos \theta}$ Ratio identities

$\qquad = \dfrac{\cos^2 \theta + \sin^2 \theta}{\sin \theta \cos \theta}$ LCD is $\sin \theta \cos \theta \sin \theta$

$\qquad = \dfrac{1}{\sin \theta \cos \theta}$ Pythagorean identity

$\qquad = \dfrac{1}{\sin \theta} \cdot \dfrac{1}{\cos \theta}$ Separate fractions

$\qquad = \csc \theta \sec \theta$ Reciprocal identities

31. $(1 - \sin \theta)(1 + \sin \theta) = 1 - \sin^2 \theta$ Multiplication

$\qquad = \cos^2 \theta$ Pythagorean identity

32. $\sin \theta (\csc \theta + \cot \theta) = \sin \theta \left(\dfrac{1}{\sin \theta} + \dfrac{\cos \theta}{\sin \theta} \right)$ Reciprocal and ratio identities

$\qquad = 1 + \cos \theta$ Multiplication

CHAPTER 2 Right Triangle Trigonometry

Problem Set 2.1

1. $a = \sqrt{c^2 - b^2}$ \hspace{1cm} Pythagorean theorem

$ = \sqrt{(5)^2 - (3)^2}$ \hspace{0.5cm} Substitute given values

$ = \sqrt{25 - 9}$ \hspace{1cm} Simplify

$ = \sqrt{16}$

$ = 4$

$\sin A = \dfrac{a}{c} = \dfrac{4}{5}$ \hspace{1.5cm} $\cot A = \dfrac{b}{a} = \dfrac{3}{4}$

$\cos A = \dfrac{b}{c} = \dfrac{3}{5}$ \hspace{1.5cm} $\sec A = \dfrac{c}{b} = \dfrac{5}{3}$

$\tan A = \dfrac{a}{b} = \dfrac{4}{3}$ \hspace{1.5cm} $\csc A = \dfrac{c}{a} = \dfrac{5}{4}$

5. $c = \sqrt{a^2 + b^2}$ \hspace{1cm} Pythagorean theorem

$ = \sqrt{(2)^2 + (\sqrt{5})^2}$ \hspace{0.3cm} Substitute given values

$ = \sqrt{4 + 5}$ \hspace{1.2cm} Simplify

$ = \sqrt{9}$

$ = 3$

$\sin A = \dfrac{a}{c} = \dfrac{2}{3}$ \hspace{1.5cm} $\cot A = \dfrac{b}{a} = \dfrac{\sqrt{5}}{2}$

$\cos A = \dfrac{b}{c} = \dfrac{\sqrt{5}}{3}$ \hspace{1.3cm} $\sec A = \dfrac{c}{b} = \dfrac{3}{\sqrt{5}}$

$\hspace{4cm} \csc A = \dfrac{c}{a} = \dfrac{3}{2}$ \hspace{0.8cm} $\tan A = \dfrac{a}{b} = \dfrac{2}{\sqrt{5}}$

9. $c = \sqrt{a^2 + b^2}$ \hspace{1cm} Pythagorean theorem

$ = \sqrt{(1)^2 + (1)^2}$ \hspace{0.5cm} Substitute given values

$ = \sqrt{1 + 1}$ \hspace{1.2cm} Simplify

$ = \sqrt{2}$

This problem continued on the next page.

$$\sin A = \frac{a}{c} = \frac{1}{\sqrt{2}} \qquad \sin B = \frac{b}{c} = \frac{1}{\sqrt{2}}$$

$$\cos A = \frac{b}{c} = \frac{1}{\sqrt{2}} \qquad \cos B = \frac{a}{c} = \frac{1}{\sqrt{2}}$$

$$\tan A = \frac{a}{b} = \frac{1}{1} = 1 \qquad \tan B = \frac{b}{a} = \frac{1}{1} = 1$$

13. $a = \sqrt{c^2 - b^2}$ Pythagorean theorem

$\qquad = \sqrt{(2x)^2 - (x)^2}$ Substitute given values

$\qquad = \sqrt{4x^2 - x^2}$ Simplify

$\qquad = \sqrt{3x^2}$

$\qquad = x\sqrt{3}$

$$\sin A = \frac{a}{c} = \frac{x\sqrt{3}}{2x} = \frac{\sqrt{3}}{2} \qquad \sin B = \frac{b}{c} = \frac{x}{2x} = \frac{1}{2}$$

$$\cos A = \frac{b}{c} = \frac{x}{2x} = \frac{1}{2} \qquad \cos B = \frac{a}{c} = \frac{x\sqrt{3}}{2x} = \frac{\sqrt{3}}{2}$$

$$\tan A = \frac{a}{b} = \frac{x\sqrt{3}}{x} = \sqrt{3} \qquad \tan B = \frac{b}{a} = \frac{x}{x\sqrt{3}} = \frac{1}{\sqrt{3}}$$

17. $\sin 10° = \cos(90° - 10°) = \cos 80°$

21. $\sin x° = \cos(90° - x°)$

25. $\csc x = \dfrac{1}{\sin x}$

$\qquad \csc 30° = \dfrac{1}{\frac{1}{2}} = 2$

$\qquad \csc 45° = \dfrac{1}{1/\sqrt{2}} = \sqrt{2}$

$\qquad \csc 60° = \dfrac{1}{\sqrt{3}/2} = \dfrac{2}{\sqrt{3}}$

$\qquad \csc 90° = \dfrac{1}{1} = 1$

29. $(2\cos 30°)^2 = [2(\frac{\sqrt{3}}{2})]^2 = (\sqrt{3})^2 = 3$

33. $\sin^2 45° - 2\sin 45° \cos 45° + \cos^2 45° = \left(\dfrac{\sqrt{2}}{2}\right)^2 - 2\left(\dfrac{\sqrt{2}}{2}\right)\left(\dfrac{\sqrt{2}}{2}\right) + \left(\dfrac{\sqrt{2}}{2}\right)^2$

$$= \dfrac{2}{4} - 2\left(\dfrac{2}{4}\right) + \dfrac{2}{4}$$

$$= \dfrac{1}{2} - 1 + \dfrac{1}{2}$$

$$= 0$$

37. $2\sin 30° = 2\left(\dfrac{1}{2}\right)$

$\qquad\qquad = 1$

41. $-3\sin 2(30°) = -3\sin 60°$

$$= -3\left(\dfrac{\sqrt{3}}{2}\right)$$

$$= -\dfrac{3\sqrt{3}}{2}$$

45. $\sec 30° = \dfrac{1}{\cos 30°}$ Reciprocal identity

$\qquad = \dfrac{1}{\sqrt{3}/2}$ Substitute exact value from Table 1

$\qquad = \dfrac{2}{\sqrt{3}}$ Division of fractions

49. $\cot 45° = \dfrac{\cos 45°}{\sin 45°}$ Ratio identity

$\qquad = \dfrac{\sqrt{2}/2}{\sqrt{2}/2}$ Substitute values from Table 1

$\qquad = 1$ Simplify

65. $\quad CH = \sqrt{(CD)^2 + (DH)^2} \qquad CF = \sqrt{(CH)^2 + (FH)^2}$

$\qquad\quad = \sqrt{5^2 + 5^2} \qquad\qquad\quad = \sqrt{(5\sqrt{2})^2 + (5)^2}$

$\qquad\quad = \sqrt{25 + 25} \qquad\qquad\quad = \sqrt{50 + 25}$

$\qquad\quad = \sqrt{50} \qquad\qquad\qquad\quad = \sqrt{75}$

$\qquad\quad = 5\sqrt{2} \qquad\qquad\qquad\quad = 5\sqrt{3}$

This problem continued on the next page.

$$\sin \theta = \frac{FH}{CF} \qquad\qquad \cos \theta = \frac{CH}{CF}$$

$$= \frac{5}{5\sqrt{3}} \qquad\qquad\qquad = \frac{5\sqrt{2}}{5\sqrt{3}}$$

$$= \frac{1}{\sqrt{3}} \qquad\qquad\qquad = \frac{\sqrt{2}}{\sqrt{3}} \text{ or } \frac{\sqrt{6}}{3}$$

69. $r = \sqrt{(x_2 - x_1)^2 + (y_2 - y_1)^2}$ Distance formula

$\quad = \sqrt{(5-2)^2 + (1-5)^2}$ Substitute given values

$\quad = \sqrt{(3)^2 + (-4)^2}$ Simplify

$\quad = \sqrt{9 + 16}$

$\quad = \sqrt{25}$

$\quad = 5$

73. If $x = 0$, then $y = 2(0) - 1$

$\qquad\qquad\qquad y = -1$

Therefore, the point $(0, -1)$ satisfies the equation.

If $x = 2$, then $y = 2(2) - 1$

$\qquad\qquad\qquad y = 4 - 1$

$\qquad\qquad\qquad y = 3$

Therefore, the point $(2, 3)$ satisfies the equation.

Plot the points $(0, -1)$ and $(2, 3)$ and draw the line through these points.

77. $-135° + 360° = 225°$

Problem Set 2.2

1. $\quad\ \ 37° \ 45'$
$\quad \underline{+\ 26° \ 24'}$
$\qquad 63° \ 69' = 64° \ 9'$ since $60' = 1°$

5. $\quad\ \ 61° \ 33'$
$\quad \underline{+\ 45° \ 16'}$
$\qquad 106° \ 49'$

9. $\quad\ \ 180° \qquad = \qquad\quad 179° \ 60'$ Change $1°$ to $60'$
$\quad \underline{-\ 120° \ 17'} \qquad\qquad \underline{-\ 120° \ 17'}$
$\qquad\qquad\qquad\qquad\qquad\qquad 59° \ 43'$

13. $\quad$ $\begin{array}{r} 70° \; 40' \\ - \; 30° \; 50' \\ \hline \end{array}$ $\quad = \quad$ $\begin{array}{r} 69° \; 100' \\ - \; 30° \; \; 50' \\ \hline 39° \; \; 50' \end{array}$ $\quad$ Change 1° to 60′

17. $16.25° = 16° + 0.25\,(60)'$

$\qquad = 16° + 15'$

$\qquad = 16° \; 15'$

21. $19.9° = 19° + 0.9(60)'$

$\qquad = 19° + 54'$

$\qquad = 19° \; 54'$

25. $62° \; 36' = 62 + \dfrac{36}{60}$

$\qquad = 62.6°$

29. $48° \; 27' = 48 + \dfrac{27}{60}$

$\qquad = 48.45°$

33. Scientific Calculator: 18 $\boxed{\cos}$

$\quad$ Graphing Calculator: $\boxed{\cos}$ $\boxed{(}$ 18 $\boxed{)}$ $\boxed{\text{ENTER}}$

$\quad$ Answer to 4 places: 0.9511

37. $\cot 31° = \dfrac{1}{\tan 31°}$

$\quad$ Scientific Calculator: 31 $\boxed{\tan}$ $\boxed{1/x}$

$\quad$ Graphing Calculator: $\boxed{\tan}$ $\boxed{(}$ 31 $\boxed{)}$ $\boxed{x^{-1}}$ $\boxed{\text{ENTER}}$

$\quad$ Answer: 1.6643

41. $\csc 14.15° = \dfrac{1}{\sin 14.15°}$

$\quad$ Scientific Calculator: 14.15 $\boxed{\sin}$ $\boxed{1/x}$

$\quad$ Graphing Calculator: $\boxed{\sin}$ $\boxed{(}$ 14.15 $\boxed{)}$ $\boxed{\text{ENTER}}$

$\quad$ Answer: 4.0906

45. $42° \; 15' = 42 + \dfrac{15}{60}$

$\qquad = 42.25°$

$\quad$ Scientific Calculator: 42.25 $\boxed{\tan}$

$\quad$ Graphing Calculator: $\boxed{\tan}$ $\boxed{(}$ 42.25 $\boxed{)}$ $\boxed{\text{ENTER}}$

$\quad$ Answer: 0.9083

49. $45° 54' = 45 + \dfrac{54}{60}$

$\qquad\qquad = 45.9°$

$\sec 45.9° = \dfrac{1}{\cos 45.9°}$

Scientific Calculator: 45.9 $\boxed{\cos}$ $\boxed{1/x}$

Graphing Calculator: $\boxed{\cos}$ $\boxed{(}$ 45.9 $\boxed{)}$ $\boxed{x^{-1}}$ $\boxed{\text{ENTER}}$

Answer: 1.4370

53.

$\sin 0° = 0$	$\cos 0° = 1$	$\tan 0° = 0$
$\sin 15° = .2588$	$\cos 15° = .9659$	$\tan 15° = .2680$
$\sin 30° = .5$	$\cos 30° = .8660$	$\tan 30° = .5774$
$\sin 45° = .7071$	$\cos 45° = .7071$	$\tan 45° = 1$
$\sin 60° = .8660$	$\cos 60° = .5$	$\tan 60° = 1.7321$
$\sin 75° = .9659$	$\cos 75° = .2588$	$\tan 75° = 3.7321$
$\sin 90° = 1$	$\cos 90° = 0$	$\tan 90°$ is undefined.

57. Scientific Calculator: 0.6873 $\boxed{\text{inv}}$ $\boxed{\tan}$

Graphing Calculator: $\boxed{\text{2nd}}$ $\boxed{\tan}$ $\boxed{(}$ 0.6873 $\boxed{)}$ $\boxed{\text{ENTER}}$

Answer: 34.5°

61. $\sec\theta = 1.0191$

$\dfrac{1}{\cos\theta} = 1.0191$

$\cos\theta = \dfrac{1}{1.0191}$

Scientific Calculator: 1 $\boxed{\div}$ 1.0191 $\boxed{=}$ $\boxed{\text{inv}}$ $\boxed{\cos}$

Graphing Calculator: $\boxed{\text{2nd}}$ $\boxed{\cos}$ $\boxed{(}$ 1 $\boxed{\div}$ 1.0191 $\boxed{)}$ $\boxed{\text{ENTER}}$

Answer: 11.1°

65. $\cot\theta = 0.6873$

$\dfrac{1}{\tan\theta} = 0.6873$

$\tan\theta = \dfrac{1}{0.6873}$

Scientific Calculator: 1 ÷ 0.6873 = inv tan

Graphing Calculator: 2nd tan (1 ÷ 0.6873) ENTER

Answer: 55.5°

69. Scientific Calculator: 0.4112 inv cos

Answer in decimal degrees is 65.719°

Convert the decimal part to minutes:

0.719 × 60 =

Graphing Calculator: 2nd cos (0.4112) 2nd MATRIX ▷DMS ENTER

To the nearest minute we have $\theta = 65° \; 43'$

73. $\sec \theta = 1.0129$

$$\frac{1}{\cos \theta} = 1.0129$$

$$\cos \theta = \frac{1}{1.0129}$$

Scientific Calculator: 1 ÷ 1.0129 = inv cos

Answer in decimal degrees is 9.154°

Convert the decimal part to minutes:

0.154 × 60 =

Graphing Calculator: 2nd cos (1 ÷ 1.0129 =)

2nd MATRIX ▷DMS ENTER

To the nearest minute we have 9° 9′

77. To calculate $\sec 34.5° = \dfrac{1}{\cos 34.5°}$:

Scientific Calculator: 34.5 cos 1/x

Graphing Calculator: cos (34.5) x⁻¹ ENTER

To calculate $\csc 55.5° = \dfrac{1}{\sin 55.5°}$:

Scientific Calculator: 55.5 sin 1/x

Graphing Calculator: sin (55.5) x⁻¹ ENTER

Both answers should be: 1.2134

81. Scientific Calculator: 37 $\boxed{\cos}$ $\boxed{x^2}$ $\boxed{+}$ 37 $\boxed{\sin}$ $\boxed{x^2}$ $\boxed{=}$

Graphing Calculator: $\boxed{\cos}$ $\boxed{(}$ 37 $\boxed{)}$ $\boxed{x^2}$ $\boxed{+}$ $\boxed{\sin}$ $\boxed{(}$ 37 $\boxed{)}$ $\boxed{x^2}$ $\boxed{\text{ENTER}}$

85. Scientific Calculator: 1.234 $\boxed{\text{inv}}$ $\boxed{\sin}$

Graphing Calculator: $\boxed{\text{2nd}}$ $\boxed{\sin}$ 1.234 $\boxed{\text{ENTER}}$

You should get an error message. The sine of an angle can never be greater than 1.

93. A point on the terminal side of an angle of 90° in standard position is (0, 1), where $x = 0$, $y = 1$, and $r = 1$.

$$\sin 90° = \frac{y}{r} = \frac{1}{1} = 1$$

$$\cos 90° = \frac{x}{r} = \frac{0}{1} = 0$$

$$\tan 90° = \frac{y}{x} = \frac{1}{0} \text{ (undefined)}$$

97. The $\sin \theta$ is positive in QI and QII.

The $\cos \theta$ is negative in QII and QIII.

Therefore, θ must lie in QII.

Problem Set 2.3

1. $\sin 42° = \dfrac{a}{15}$ Sine relationship

$a = 15 \sin 42°$ Subtract α from both sides

$= 15(0.6691)$ Substitute value for $\sin 42°$

$= 10$ ft   Answer rounded to 2 significant digits

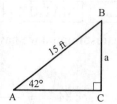

5. $\cos 24.5° = \dfrac{a}{2.34}$ Cosine relationship

$a = 2.34 \cos 24.5°$ Multiply both sides by 2.34

$= 2.34(0.9099)$ Substitute value for $\cos 24.5°$

$= 2.13$ ft   Answer rounded to 3 significant digits

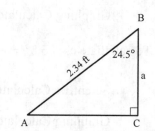

9. $\tan A = \dfrac{16}{26}$ Tangent relationship

  $= 0.6153$ Divide 16 by 26

  $A = 32°$ Answer rounded to the nearest degree

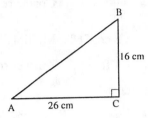

13. $\cos B = \dfrac{23.32}{45.54}$ Cosine relationship

  $= 0.5120$ Divide 23.32 by 45.54

  $b = 59.20°$ Answer rounded to nearest hundredth of a degree

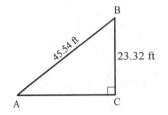

17.   First, we find $\angle B$: $\angle B = 90° - \angle A$

  $= 90° - 32.6°$

  $= 57.4°$

Next, we find side c:

$\sin 32.6° = \dfrac{43.4}{c}$ Sine relationship

  $c = \dfrac{43.4}{\sin 32.6°}$ Multiply both sides by c then divide by 32.6°

  $= 80.6$ in   Answer rounded to 3 significant digits

Last, we find side b:

$\tan 57.4° = \dfrac{b}{43.4}$ Tangent relationship

  $b = 43.4 \tan 57.4°$ Multiply both sides by 43.4

  $= 67.9$ in   Answer rounded to 3 significant digits

21.   First, we find $\angle A$: $\angle A = 90° - 76°$

  $= 14°$

Next, we find side a:

This problem continued on the next page.

$$\cos 76° = \frac{a}{5.8}$$ Cosine relationship

$$a = 5.8 \cos 76°$$ Multiply both sides by 5.8

$$= 1.4 \text{ ft}$$ Answer rounded to 2 significant digits

Last, we find side b:

$$\sin 76° = \frac{b}{5.8}$$ Sine relationship

$$b = 5.8 \sin 76°$$ Multiply both sides by 5.8

$$= 5.6 \text{ ft}$$ Answer rounded to 2 significant digits

25. First, we find $\angle A$:

$$\angle A = 90° - 23.45°$$

$$= 66.55°$$

Next, we find side b:

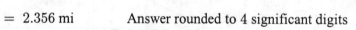

$$\tan 23.45° = \frac{b}{5.432}$$ Tangent relationship

$$b = 5.432 \tan 23.45°$$ Multiply both sides by 5.432

$$= 2.356 \text{ mi}$$ Answer rounded to 4 significant digits

Last, we find side c:

$$\cos 23.45° = \frac{5.432}{c}$$ Cosine relationship

$$c = \frac{5.432}{\cos 23.45°}$$ Multiply both sides by c and then divide by $\cos 23.45°$

$$= 5.921 \text{ mi}$$ Answer rounded to 4 significant digits

29. First, we find $\angle A$:

$$\sin A = \frac{2.75}{4.05}$$ Sine relationship

$$= 0.6790$$ Divide 2.75 by 4.05

$$A = 42.8°$$ Answer rounded to the

nearest tenth of a degree

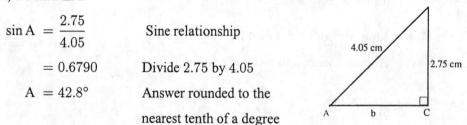

Next, we find $\angle B$:

$$\angle B = 90° - 42.8°$$

$$= 47.2°$$

Last, we find side b:

$b^2 + (2.75)^2 = (4.05)^2$ Pythagorean theorem

$b^2 + 7.5625 = 16.4025$ Simplify

$b^2 = 8.84$ Subtract 7.5625 from both sides

$b = \pm 2.97$ Take square root of both sides

$= 2.97$ cm b must be positive

33.   Using $\triangle$ BCD, we find BD:

$\sin 30° = \dfrac{BD}{6}$ Sine relationship

$BD = 6 \sin 30°$ Multiply both sides by 6

$= 3$ Exact answer

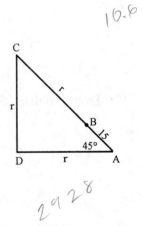

Next, we find $\angle$B: $\angle B = 90° - 42.8°$

$\sin A = \dfrac{3}{4}$ Sine relationship

$= 0.75$ Divide 3 by 4

$A = 49°$ Answer rounded to the nearest degree

10.8

37. $\sin 45° = \dfrac{r}{r + 15}$

$r = (r + 15)\sin 45°$ Sine relationship

$r = r \sin 45° + 15 \sin 45°$ Multiply both sides by $r + 15$

$r - r \sin 45° = 15 \sin 45°$ Use distributive property

$r(1 - \sin 45°) = 15 \sin 45°$ Subtract $r \sin 45°$ from both sides

$r = \dfrac{15 \sin 45°}{1 - \sin 45°}$ Divide both sides by $1 - \sin 45°$

$= 36$ Answer rounded to 2 significant digits

2928

41.   Using $\triangle$ ABC, we find side x:

$\sin 41° = \dfrac{x}{32}$ Sine relationship

$x = 32 \sin 41°$ Multiply both sides by 32

$= 21$ Answer rounded to 2 significant digits

Next, using $\triangle$ ABD, we find $\angle$ABD:

$\tan \angle ABD = \dfrac{h}{x}$ Tangent relationship

$$= \frac{19}{21} \qquad \text{Substitute given values}$$

$$= 0.9047 \qquad \text{Divide 19 by 21}$$

$$\angle \text{ABD} = 42° \qquad \text{Answer rounded to the nearest degree}$$

45. Using $\triangle$ ABC, we find side h:

$$\sin 41° = \frac{h}{28} \qquad \text{Sine relationship}$$

$$h = 28 \sin 41° \qquad \text{Multiply both sides by 28}$$

$$= 18 \qquad \text{Answer rounded to 2 significant digits}$$

Next, using $\triangle$ BCD, we find side x:

$$\tan 58° = \frac{h}{x} \qquad \text{Tangent relationship}$$

$$\tan 58° = \frac{18}{x} \qquad \text{Substitute value found for } h$$

$$x = \frac{18}{\tan 58°} \qquad \text{Multiply both sides by } x \text{ and divide by } \tan 58°$$

$$= 11 \qquad \text{Answer rounded to 2 significant digits}$$

49. From Problem 65 in Problem Set 2.1, we found that $\sin \theta = \dfrac{1}{\sqrt{3}}$

$$= 0.5774$$

$$\theta = 35.3°$$

53.

$$\cos 30° = \frac{x}{125} = \frac{139 - h}{125}$$

$$125 \cos 30° = 139 - h$$

$$\cos 30° = \frac{139 - h}{125}$$

$$h = 139 - 125 \cos 30° \qquad \text{Solve for } h$$

$$= 139 - 108.25$$

$$= 30.7 \text{ ft} \qquad \text{Round to nearest tenth}$$

57. $r = 98.5$

   a. $h = 12 + 98.5 + x$

$$\cos 60° = \frac{x}{98.5}$$

$$x = 98.5 \cos 60°$$

$$= 49.25$$

$$h = 12 + 98.5 + 49.25$$

$$= 159.8 \text{ ft}$$

   b. $h = 12 + 98.5 + x$

$$\cos 30° = \frac{x}{98.5}$$

$$x = 98.5 \cos 30°$$

$$= 85.3$$

$$h = 12 + 98.5 + 85.3$$

$$= 195.8 \text{ ft}$$

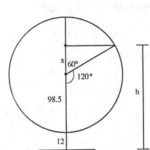

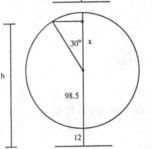

   c. $r + 12 = 98.5 + 12 = 110.5$

$$h = 110.5 - x$$

$$\cos 45° = \frac{x}{98.5}$$

$$x = 98.5 \cos 45°$$

$$= 69.7$$

$$h = 110.5 - 69.7$$

$$= 40.8 \text{ ft}$$

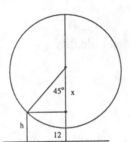

61. $\cos \theta = \sqrt{1 - \sin^2 \theta}$ θ is in QI

$$= \sqrt{1 - \left(\frac{1}{3}\right)^2}$$

$$= \sqrt{1 - \frac{1}{9}}$$

$$= \sqrt{\frac{8}{9}}$$

$$= \frac{2\sqrt{2}}{3}$$

65. $\cos \theta = -\sqrt{1 - \sin^2 \theta}$ Pythagorean identity, θ in QII

$\qquad = -\sqrt{1 - \left(\dfrac{\sqrt{3}}{2}\right)^2}$ Substitute given value

$\qquad = -\sqrt{1 - \dfrac{3}{4}}$ Simplify

$\qquad = -\sqrt{\dfrac{1}{4}}$

$\qquad = -\dfrac{1}{2}$

$\tan \theta = \dfrac{\sin \theta}{\cos \theta}$ Ratio identity

$\qquad = \dfrac{\sqrt{3}/2}{-1/2}$ Substitute given values

$\qquad = -\sqrt{3}$ Simplify

$\csc \theta = \dfrac{1}{\sin \theta}$ Reciprocal identity

$\qquad = \dfrac{1}{\sqrt{3}/2}$

$\qquad = \dfrac{2}{\sqrt{3}}$

$\sec \theta = \dfrac{1}{\cos \theta}$ Reciprocal identity

$\qquad = \dfrac{1}{-1/2}$

$\qquad = -2$

$\cot \theta = \dfrac{1}{\tan \theta}$ Reciprocal identity

$\qquad = \dfrac{1}{-\sqrt{3}}$

$\qquad = -\dfrac{1}{\sqrt{3}}$

Problem Set 2.4

1. To find the height, h, we can use the Pythagorean theorem:

$$h^2 + (15)^2 = (42)^2$$
$$h^2 + 225 = 1{,}764$$
$$h^2 = 1{,}539$$
$$h = \pm\sqrt{1{,}539}$$
$$= 39 \text{ cm}$$

To find angle θ, we can use the cosine ratio:

$$\cos\theta = \frac{15}{42}$$
$$= 0.3571$$
$$\theta = 69°$$

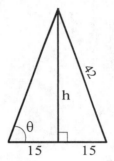

5. To find the length of the escalator, x, we use the sine ratio:

$$\sin 33° = \frac{21}{x}$$
$$x = \frac{21}{\sin 33°}$$
$$= \frac{21}{0.5446}$$
$$= 39 \text{ ft}$$

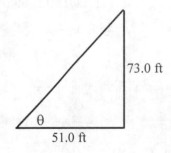

9. We use the tangent ratio to find the angle of elevation to the sun, θ:

$$\tan\theta = \frac{73.0}{51.0}$$
$$= 1.4313$$
$$\theta = 55.1°$$

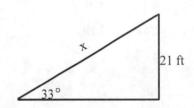

13. $\tan 65° = \dfrac{x}{18}$

$$x = 18\tan 65°$$
$$= 18(2.1445)$$
$$= 39 \text{ mi}$$

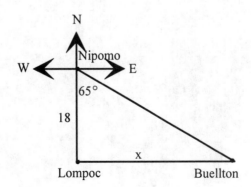

21.

$$\tan 59° = \frac{10}{y}$$

$$y = \frac{10}{\tan 59°}$$

$$= \frac{10}{1.6643}$$

$$= 6.0$$

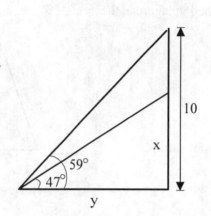

$$\tan 47° = \frac{10}{\tan 59°}$$

$$x = y \tan 47°$$

$$= 6.0(1.0724)$$

$$= 6.4 \text{ ft}$$

25.

$$\tan 86.6° = \frac{x}{24.8}$$

$$x = 24.8 \tan 86.6°$$

$$= 24.8(16.8319)$$

$$= 417.431$$

$$\tan 10.7° = \frac{h}{x}$$

$$h = x \tan 10.7°$$

$$= (417.431)(0.18895)$$

$$= 78.9 \text{ ft}$$

29.

$$\sin 76.6° = \frac{r}{r + 112}$$

$$r = (r + 112)\sin 76.6°$$

$$r = r \sin 76.6° + 112 \sin 76.6°$$

$$r - r \sin 76.6° = 112 \sin 76.6°$$

$$r(1 - \sin 76.6°) = 112 \sin 76.6°$$

$$r = \frac{112 \sin 76.6°}{1 - \sin 76.6°}$$

$$= \frac{112(0.9728)}{1 - 0.9728}$$

$$= \frac{108.9509}{0.02722}$$

$$= 4,000 \text{ mi}$$

29.
$$\sin 76.6° = \frac{r}{r+112}$$
$$r = (r+112)\sin 76.6°$$
$$r = r \sin 76.6° + 112 \sin 76.6°$$
$$r - r \sin 76.6° = 112 \sin 76.6°$$
$$r(1 - \sin 76.6°) = 112 \sin 76.6°$$
$$r = \frac{112 \sin 76.6°}{1 - \sin 76.6°}$$
$$= \frac{112(0.9728)}{1 - 0.9728}$$
$$= \frac{108.9509}{0.02722}$$
$$= 4,000 \text{ mi}$$

33.

$$\tan \theta_1 = \frac{1}{1} \qquad \tan \theta_2 = \frac{1}{\sqrt{2}} \qquad \tan \theta_3 = \frac{1}{\sqrt{3}}$$
$$= 1 \qquad\qquad\quad = 0.7071 \qquad\quad = 0.5774$$
$$\theta_1 = 45° \qquad\quad\ \theta_2 = 35.26° \qquad \theta_3 = 30°$$

37.

$$\sin \theta \cot \theta = \sin \theta \cdot \frac{\cos \theta}{\sin \theta} \qquad \text{Ratio identity}$$
$$= \frac{\sin \theta \cos \theta}{\sin \theta} \qquad \text{Multiplication of fractions}$$
$$= \cos \theta \qquad \text{Division of common factor}$$

41.

$$\sec \theta - \cos \theta = \frac{1}{\cos \theta} - \cos \theta \qquad \text{Reciprocal identity}$$
$$= \frac{1}{\cos \theta} - \cos \theta \cdot \frac{\cos \theta}{\cos \theta} \qquad \text{LCD is } \cos \theta$$
$$= \frac{1 - \cos^2 \theta}{\cos \theta} \qquad \text{Subtraction of fractions}$$
$$= \frac{\sin^2 \theta}{\cos \theta} \qquad \text{Pythagorean identity}$$

Problem Set 2.5

13. To find the magnitude of $\vec{V}$, we use the Pythagorean theorem:

$$|\vec{V}| = \sqrt{(3.25)^2 + (12.0)^2}$$

$$= \sqrt{10.5625 + 144}$$

$$= \sqrt{154.5625}$$

$$= 12.4 \text{ mph}$$

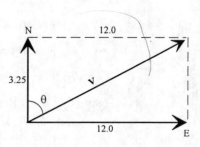

To find the direction of the boat, we find θ using the tangent ratio:

$$\tan \theta = \frac{12.0}{3.25}$$

$$\tan \theta = 3.6923$$

$$\theta = 74.8°$$

Therefore, the true course of the boat is 12.4 miles per hour at N 74.8° E.

17. We want to find the magnitude of vector, $\vec{V}$, which we can do using the tangent ratio:

$$\tan 78° = \frac{12}{|\vec{V}|}$$

$$|\vec{V}| = \frac{12}{\tan 78°}$$

$$= \frac{12}{4.7046}$$

$$= 2.6 \text{ mph}$$

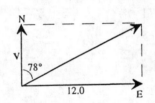

21. To find the distance, x, the plane has flown off its course, we can use the sine ratio:

$$\sin 3° = \frac{x}{130}$$

$$x = 130 \sin 3°$$

$$= 6.8 \text{ miles}$$

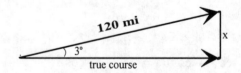

25.

$$|\vec{V_x}| = |\vec{V}| \cos \theta \qquad\qquad |\vec{V_y}| = |\vec{V}| \sin \theta$$

$$= 420 \cos 36° 10' \qquad\qquad = 420 \sin 36° 10'$$

$$= 420 \cos 36.17° \qquad\qquad = 420 \sin 36.17°$$

$$= 420(0.8073) \qquad\qquad = 420(0.5901)$$

$$= 339 \qquad\qquad\qquad\quad = 248$$

29.
$$\vec{|V|} = \sqrt{\vec{|V_x|}^2 + \vec{|V_y|}^2}$$
$$= \sqrt{(35.0)^2 + (26.0)^2}$$
$$= \sqrt{1{,}225 + 676}$$
$$= \sqrt{1{,}901}$$
$$= 43.6$$

33.
$$\vec{|V_x|} = \vec{|V|} \cos\theta \qquad\qquad \vec{|V_y|} = \vec{|V|} \sin\theta$$
$$= 1{,}200 \cos 45° \qquad\qquad = 1{,}200 \sin 45°$$
$$= 1{,}200(0.7071) \qquad\qquad = 1{,}200(0.7071)$$
$$= 850 \text{ feet per second} \qquad = 850 \text{ feet per second}$$

37. We are given that $\vec{|V_x|} = 35.0$ and $\vec{|V_y|} = 15.0$.

$$\vec{|V|} = \sqrt{\vec{|V_x|}^2 + \vec{|V_y|}^2}$$
$$= \sqrt{(35.0)^2 + (15.0)^2}$$
$$= \sqrt{1{,}225 + 225}$$
$$= \sqrt{1{,}450}$$
$$= 38.1 \text{ feet per second}$$

$$\tan\theta = \frac{\vec{|V_y|}}{\vec{|V_x|}}$$
$$= \frac{15.0}{35.0}$$
$$= 0.4285$$
$$\theta = 23.2°$$

Therefore, the velocity of the arrow is 38.1 feet per second at an angle inclination of 23.2°.

41. To find the total distance traveled north, we must find the sum of $\vec{|V_y|}$ and $\vec{|W_y|}$ and to find the total distance traveled east, we must find the sum of $\vec{|V_x|}$ and $\vec{|W_x|}$.

We are given that $\vec{|V|}$ is 175 mi. at an angle of inclination of $90° - 18°$ or $72°$ and also that $\vec{|W|}$ is 120 mi. at an angle of inclination of $90° - 49°$ or $41°$.

$$|\vec{V}_x| = |\vec{V}| \cos\theta_1 \qquad\qquad |\vec{V}_y| = |\vec{V}| \sin\theta_1$$
$$= 175 \cos 72° \qquad\qquad\quad = 175 \sin 72°$$
$$= 175(0.3090) \qquad\qquad = 175(0.9510)$$
$$= 54 \text{ mi} \qquad\qquad\qquad = 166 \text{ mi}$$

$$|\vec{W}_x| = |\vec{W}| \cos\theta_2 \qquad\qquad |\vec{W}_y| = |\vec{W}| \sin\theta_2$$
$$= 120 \cos 41° \qquad\qquad\quad = 120 \sin 41°$$
$$= 120(0.7547) \qquad\qquad = 120(0.6560)$$
$$= 91 \text{ mi} \qquad\qquad\qquad = 79 \text{ mi}$$

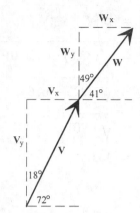

Therefore, the total distance north is $|\vec{V}_y| + |\vec{W}_y| = 166 + 79 = 245$ miles

and the total distance east is $|\vec{V}_x| + |\vec{W}_x| = 54 + 91 = 145$ miles

45. We are given that $|\vec{W}| = 10$

$$\cos 15° = \frac{|\vec{N}|}{|\vec{W}|} \qquad\qquad \sin 15° = \frac{|\vec{F}|}{|\vec{W}|}$$
$$|\vec{N}| = |\vec{W}| \cos 15° \qquad\qquad |\vec{F}| = |\vec{W}| \sin 15°$$
$$= 10(0.9659) \qquad\qquad\quad = 10(0.2588)$$
$$= 9.7 \text{ pounds} \qquad\qquad = 2.6 \text{ pounds}$$

49. $(x, y) = (-1, 1)$

$x = -1, y = 1$ and $r = \sqrt{2}$

$$\sin 135° = \frac{y}{r} = \frac{1}{\sqrt{2}}$$

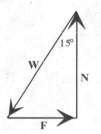

$$\cos 135° = \frac{x}{r} = -\frac{1}{\sqrt{2}}$$

$$\tan 135° = \frac{y}{x} = \frac{1}{-1} = -1$$

53. $\sin\theta = \dfrac{y}{r} = \dfrac{-4}{5} = \dfrac{-8}{10}$

$y = -8$ and $r = 10$

$$x^2 + y^2 = r^2$$
$$x^2 + (-8)^2 = 10^2$$
$$x^2 + 64 = 100$$
$$x^2 = 36$$
$$x = \pm 6$$

Chapter 2 Test

1. $c = \sqrt{a^2 + b^2}$

$\quad = \sqrt{(1)^2 + (2)^2}$

$\quad = \sqrt{1 + 4}$

$\quad = \sqrt{5}$

$\sin A = \dfrac{a}{c} = \dfrac{1}{\sqrt{5}}$ $\qquad\qquad$ $\sin B = \dfrac{b}{c} = \dfrac{2}{\sqrt{5}}$

$\cos A = \dfrac{b}{c} = \dfrac{2}{\sqrt{5}}$ $\qquad\qquad$ $\cos B = \dfrac{a}{c} = \dfrac{1}{\sqrt{5}}$

$\tan A = \dfrac{a}{b} = \dfrac{1}{2}$ $\qquad\qquad$ $\tan B = \dfrac{b}{a} = \dfrac{2}{1} = 2$

2. $a = \sqrt{c^2 - b^2}$

$\quad = \sqrt{6^2 - 3^2}$

$\quad = \sqrt{36 - 9}$

$\quad = \sqrt{27}$

$\quad = 3\sqrt{3}$

$\sin A = \dfrac{a}{c} = \dfrac{3\sqrt{3}}{6} = \dfrac{\sqrt{3}}{2}$ $\qquad$ $\sin B = \dfrac{b}{c} = \dfrac{3}{6} = \dfrac{1}{2}$

$\cos A = \dfrac{b}{c} = \dfrac{3}{6} = \dfrac{1}{2}$ $\qquad$ $\cos B = \dfrac{a}{c} = \dfrac{3\sqrt{3}}{6} = \dfrac{\sqrt{3}}{2}$

$\tan A = \dfrac{a}{b} = \dfrac{3\sqrt{3}}{3} = \sqrt{3}$ $\qquad$ $\tan B = \dfrac{b}{a} = \dfrac{3}{3\sqrt{3}} = \dfrac{1}{\sqrt{3}}$

3. $b = \sqrt{c^2 - a^2}$

$\quad = \sqrt{5^2 - 3^2}$

$\quad = \sqrt{25 - 9}$

$\quad = \sqrt{16} = 4$

$\sin A = \dfrac{a}{c} = \dfrac{3}{5}$ $\qquad\qquad$ $\sin B = \dfrac{b}{c} = \dfrac{4}{5}$

$\cos A = \dfrac{b}{c} = \dfrac{4}{5}$ $\qquad\qquad$ $\cos B = \dfrac{a}{c} = \dfrac{3}{5}$

$\tan A = \dfrac{a}{b} = \dfrac{3}{4}$ $\qquad\qquad$ $\tan B = \dfrac{b}{a} = \dfrac{4}{3}$

4.
$$c = \sqrt{a^2 + b^2}$$
$$= \sqrt{5^2 + 12^2}$$
$$= \sqrt{25 + 144}$$
$$= \sqrt{169}$$
$$= 13$$

$$\sin A = \frac{a}{c} = \frac{5}{13} \qquad\qquad \sin B = \frac{b}{c} = \frac{12}{13}$$

$$\cos A = \frac{b}{c} = \frac{12}{13} \qquad\qquad \cos B = \frac{a}{c} = \frac{5}{13}$$

$$\tan A = \frac{a}{b} = \frac{5}{12} \qquad\qquad \tan B = \frac{b}{a} = \frac{12}{5}$$

5.
$$\sin 14° = \cos(90° - 14°)$$
$$= \cos 76°$$

6.
$$\csc 73° = \sec(90° - 73°)$$
$$= \sec 17°$$

7.
$$\sin^2 45° + \cos^2 30° = \left(\frac{1}{\sqrt{2}}\right)^2 + \left(\frac{\sqrt{3}}{2}\right)^2 = \frac{1}{2} + \frac{3}{4} = \frac{5}{4}$$

8.
$$\tan 45° + \cot 45° = 1 + 1$$
$$= 2$$

9.
$$\sin^2 60° - \cos^2 30° = \left(\frac{\sqrt{3}}{2}\right)^2 - \left(\frac{\sqrt{3}}{2}\right)^2 = 0$$

10.
$$\frac{1}{\sec 30°} = \cos 30°$$
$$= \frac{\sqrt{3}}{2}$$

11.
$$\begin{array}{r} 48° \ 18' \\ + \ 24° \ 52' \\ \hline 72° \ 70' = 73° \ 10' \end{array}$$

12.
$$\begin{array}{rl} 25° \ 15' = & 24° \ 75' \qquad 1° = 60' \\ - \ 15° \ 32' & \underline{15° \ 32'} \\ & \ \ 9° \ 43' \end{array}$$

13.
$$73.2° = 73° + 0.2(60)'$$
$$= 73° \ 12'$$

14. $16.45° = 16° + 0.45(60)'$

$\qquad = 16° 27'$

15. $2° 48' = 2° + (\dfrac{48}{60})°$

$\qquad = 2° + 0.8°$

$\qquad = 2.8°$

16. $79° 30' = 79° + (\dfrac{30}{60})°$

$\qquad = 79° + 0.5°$

$\qquad = 79.5°$

17. $\sin 24° 20' = \sin 24.33°$

$\qquad = 0.4120$

18. $\cos 37.8° = 0.7902$

19. $\tan 63° 50' = \tan 63.833°$

$\qquad = 2.0353$

20. $\cot 71° 20' = \cot 71.333°$

$\qquad = 0.3378$

27. $c = \sqrt{a^2 + b^2}$ $\qquad \tan A = \dfrac{68.0}{104}$

$\qquad = \sqrt{(68.0)^2 + (104)^2}$ $\qquad = 0.6538$

$\qquad = \sqrt{4{,}624 + 10{,}816}$ $\qquad A = 33.2°$

$\qquad = \sqrt{15{,}440}$

$\qquad = 124$ $\qquad B = 90° - 33.2°$

$\qquad\qquad\qquad = 56.8°$

28. $b = \sqrt{c^2 - a^2}$ $\qquad \sin A = \dfrac{24.3}{48.1}$

$\qquad = \sqrt{(48.1)^2 - (24.3)^2}$ $\qquad = 0.5052$

$\qquad = \sqrt{2{,}313.61 - 590.49}$ $\qquad A = 30.3°$

$\qquad = \sqrt{1{,}723.12}$ $\qquad B = 90° - 30.3°$

$\qquad = 41.5$ $\qquad\qquad = 59.7°$

29. $A = 90° - 24.9°$

$\qquad = 65.1°$

$\sin 24.9° = \dfrac{305}{c}$ $\qquad\qquad$ $\tan 65.1° = \dfrac{a}{305}$

$\qquad\quad c = \dfrac{305}{\sin 24.9°}$ $\qquad\qquad\qquad a = 305 \tan 65.1°$

$\qquad\qquad\qquad\qquad\qquad\qquad\qquad\qquad = 305(2.1543)$

$\qquad\quad\ = \dfrac{305}{0.4210}$ $\qquad\qquad\qquad\qquad\quad\ = 657$

$\qquad\quad\ = 724$

30. $b = 90° - 35° \, 30'$

$\qquad = 54° \, 30'$

$\sin 35° \, 30' = \dfrac{a}{0.462}$ $\qquad\qquad$ $\sin 54° \, 30' = \dfrac{b}{0.462}$

$\qquad\qquad a = 0.462 \sin 35.5°$ $\qquad\qquad\qquad b = 0.462 \sin 54.5°$

$\qquad\qquad\ \ = 0.462(0.5807)$ $\qquad\qquad\qquad\ \ = 0.462(0.8141)$

$\qquad\qquad\ \ = 0.268$ $\qquad\qquad\qquad\qquad\ \ = 0.376$

31. $\sin 17° = \dfrac{25}{x}$

$\qquad\quad x = \dfrac{25}{\sin 17°}$

$\qquad\qquad = \dfrac{25}{0.2924}$

$\qquad\qquad = 86 \text{ cm}$

32. $\tan 75° \, 30' = \dfrac{x}{1.5}$

$\qquad\qquad x = 1.5 \tan 75.5°$

$\qquad\qquad\ \ = 1.5(3.8667)$

$\qquad\qquad\ \ = 5.8 \text{ ft}$

33. $\tan 43° = \dfrac{35}{x}$

$x = \dfrac{35}{\tan 43°}$

$= \dfrac{35}{0.9325}$

$= 37.5 \text{ ft}$

$\tan 47° = \dfrac{35}{y}$

$y = \dfrac{35}{\tan 47°}$

$= \dfrac{35}{1.0724}$

$= 32.6 \text{ ft}$

$x + y = 70 \text{ feet (rounded to 2 significant digits)}$

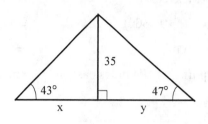

34. $|\vec{V}_x| = 5.0 \cos 30°$　　　　$|\vec{V}_y| = 5.0 \sin 30°$

$\quad = 5.0(0.8660)$　　　　　$= 5.0(0.5)$

$\quad = 4.3$　　　　　　　　　$= 2.5$

35. $\tan \theta = \dfrac{|\vec{V}_y|}{|\vec{V}_x|}$

$= \dfrac{31}{11}$

$= 2.8182$

$\theta = 70°$

36. $|\vec{V}_x| = 800 \cos 62°$　　　　$|\vec{V}_y| = 800 \sin 62°$

$\quad = 800(0.4695)$　　　　　$= 800(0.8829)$

$\quad = 380 \text{ ft}$　　　　　　　$= 710 \text{ ft}$

37.
$$\theta = -30° + 360° = 330°$$

$$|\vec{V_x}| = |\vec{V}|\cos\theta \qquad\qquad |\vec{V_y}| = |\vec{V}|\sin\theta$$
$$= 120\cos 330° \qquad\qquad = 120\sin 330°$$
$$= 120(0.8660) \qquad\qquad = 120(-0.5)$$
$$= 104 \qquad\qquad\qquad = -60$$

The ship travels 104 miles east and 60 miles south.

38.
$$x = \sqrt{(45.5)^2 + (245)^2}$$
$$x = \sqrt{2,070.25 + 60,025}$$
$$= \sqrt{62,095.25}$$
$$= 249 \text{ mph}$$

$$\tan\theta = \frac{45.5}{245}$$
$$= 0.1857$$
$$\theta = 10.5°$$

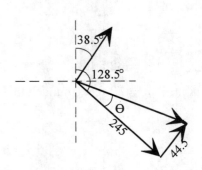

Bearing is $128.5° - 10.5° = 118.0°$ clockwise from due north.

39.
$$\tan 25.5° = \frac{|\vec{H}|}{|\vec{W}|}$$
$$= \frac{|\vec{H}|}{95.5}$$
$$|\vec{H}| = 95.5\tan 25.5°$$
$$= 95.5(0.4770)$$
$$= 45.6 \text{ pounds}$$

40.
$$\sin 8.5° = \frac{|\vec{F}|}{|\vec{W}|}$$
$$= \frac{|\vec{F}|}{58.0}$$

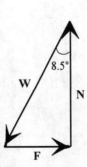

$$\frac{|\vec{F}|}{58.0} = 58.0\sin 8.5°$$
$$= 58.0(0.1478)$$
$$= 8.6 \text{ pounds}$$

CHAPTER 3 Radian Measure

Problem Set 3.1

1. $210° − 180° = 30°$

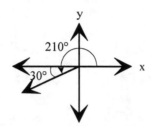

5. $360° − 311.7° = 48.3°$

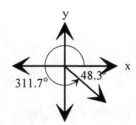

9. $−300° + 360° = 60°$

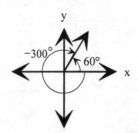

13. $\widehat{\theta} = 225° − 180°$

$= 45°$

Since θ terminates in QIII, $\cos \theta$ is negative.

$\cos 225° = −\cos 45°$

$$= −\frac{1}{\sqrt{2}}$$

17. $\widehat{\theta} = 180° − 135°$

$= 45°$

Since θ terminates in QII, $\tan \theta$ is negative.

$\tan 135° = −\tan 45°$

$= −1$

21. $\widehat{\theta} = 360° - 330°$

$\qquad = 30°$

Since θ terminates in QIV, $\csc \theta$ is negative

$\csc 330° = -\csc 30°$

$\qquad = -2$

25. $\widehat{\theta} = 390° - 360°$

$\qquad = 30°$

Since θ terminates in QI, $\sin \theta$ is positive.

$\sin 390° = \sin 30°$

$\qquad = \dfrac{1}{2}$

49. First, we find $\widehat{\theta}$:

$\sin \widehat{\theta} = 0.3090$

$\widehat{\theta} = 18.0°$ 	Rounded to the nearest tenth of a degree

Scientific Calculator: 0.3090 | inv | | sin |

Graphing Calculator: | 2nd | | sin | | (| 0.3090 |) | | ENTER |

Since θ is in QIII,

$\theta = 180° + 18.0°$

$\qquad = 198.0°$

53. First, we find $\widehat{\theta}$:

$\tan \widehat{\theta} = 0.5890$

$\widehat{\theta} = 30.5°$ 	Rounded to the nearest tenth of a degree

Scientific Calculator: 0.5890 | inv | | tan |

Graphing Calculator: | 2nd | | tan | | (| 0.5890 |) | | ENTER |

Since θ is in QIII,

$\theta = 180° + 30.5°$

$\qquad = 210.5°$

57. First, we find $\widehat{\theta}$:

$$\sin\widehat{\theta} = 0.9652$$

$$\widehat{\theta} = 74.8° \qquad \text{Rounded to the nearest tenth of a degree}$$

Scientific Calculator: 0.9652 | inv | sin |

Graphing Calculator: | 2nd | sin | (| 0.9652 |) | ENTER |

Since θ is in QII,

$$\theta = 180° - 74.8°$$

$$= 105.2°$$

61. First, we find $\widehat{\theta}$:

$$\csc\widehat{\theta} = 2.4957$$

$$\sin\widehat{\theta} = \frac{1}{2.4957}$$

$$\widehat{\theta} = 23.6° \qquad \text{Rounded to the nearest tenth of a degree}$$

Scientific Calculator: 2.4957 | 1/x | inv | sin |

Graphing Calculator: | 2nd | sin | (| 1 | ÷ | 2.4957 |) | ENTER |

Since θ is in QII,

$$\theta = 180° - 23.6°$$

$$= 156.4°$$

65. First, we find $\widehat{\theta}$:

$$\sec\widehat{\theta} = 1.7876$$

$$\cos\widehat{\theta} = \frac{1}{1.7876}$$

$$\widehat{\theta} = 56.0° \qquad \text{Rounded to the nearest tenth of a degree}$$

Scientific Calculator: 1.7876 | 1/x | inv | cos |

Graphing Calculator: | 2nd | cos | (| 1 | ÷ | 1.7876 |) | ENTER |

Since θ is in QIII,

$$\theta = 180° + 56.0°$$

$$= 236.0°$$

69. First, we find $\widehat{\theta}$:

$$\cos\widehat{\theta} \;=\; \frac{1}{\sqrt{2}} \qquad \text{This is an exact value}$$

$$\widehat{\theta} \;=\; 45°$$

Since θ is in QII,

$$\theta \;=\; 180° - 45°$$

$$=\; 135°$$

73. First, we find $\widehat{\theta}$:

$$\tan\widehat{\theta} \;=\; \sqrt{3} \qquad \text{This is an exact value}$$

$$\widehat{\theta} \;=\; 60°$$

Since θ is in QIII,

$$\theta \;=\; 180° + 60°$$

$$=\; 240°$$

77. First, we find $\widehat{\theta}$:

$$\csc\widehat{\theta} \;=\; \sqrt{2}$$

$$\sin\widehat{\theta} \;=\; \frac{1}{\sqrt{2}} \qquad \text{This is an exact value}$$

$$\widehat{\theta} \;=\; 45°$$

Since θ is in QII,

$$\theta \;=\; 180° - 45° \;=\; 135°$$

81. The complement of $70°$ is $20°$ because $70° + 20° = 90°$.

The supplement of $70°$ is $110°$ because $70° + 110° = 180°$.

85. The side opposite the $30°$ angle is one-half of the longest side, or $\frac{1}{2} \cdot 10 = 5$.

The side opposite the $60°$ angle is $\sqrt{3}$ times the shortest side, or $5\sqrt{3}$.

89.
$$\sin^2 45° + \cos^2 45° \;=\; \left(\frac{1}{\sqrt{2}}\right)^2 + \left(\frac{1}{\sqrt{2}}\right)^2$$

$$=\; \frac{1}{2} + \frac{1}{2}$$

$$=\; 1$$

Problem Set 3.2

1. $\theta = \dfrac{s}{r}$ Definition of radian measure

    $= \dfrac{9 \text{ cm}}{3 \text{ cm}}$ Substitute given values

    $= 3$ radians   Divide

5. $\theta = \dfrac{s}{r}$ Definition of radian measure

    $= \dfrac{12\pi \text{ inches}}{4 \text{ inches}}$ Substitute given values

    $= 3\pi$ radians   Divide

9. $\theta = \dfrac{s}{r}$ Definition of radian measure

    $= \dfrac{450}{4000}$ Substitute given values

    $= \dfrac{9}{80}$ or 0.1125 radians   Divide

13. (b) $90° = 90(\frac{\pi}{180})$

        $= \frac{\pi}{2}$ radians

    (c) Reference angle is itself:

      $90° = \frac{\pi}{2}$ radians

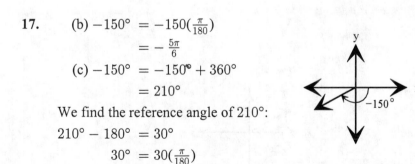

17. (b) $-150° = -150(\frac{\pi}{180})$

           $= -\frac{5\pi}{6}$

    (c) $-150° = -150° + 360°$

            $= 210°$

We find the reference angle of $210°$:

$210° - 180° = 30°$

       $30° = 30(\frac{\pi}{180})$

           $= \frac{\pi}{6}$ radians

21.   (b) $-135° = -135(\frac{\pi}{180})$

$= -\frac{3\pi}{4}$ radians

(c) $-135° = -135° + 360°$

$= 225°$

We find the reference angle of 225°:

$225° - 180° = 45°$

$45° = 45(\frac{\pi}{180})$

$= \frac{\pi}{4}$ radians

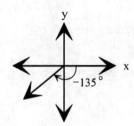

25.   1 minute $= \frac{1}{60}$ degree

$= \frac{1}{60}\left(\frac{\pi}{180}\right)$

$= \frac{\pi}{10,800}$

$= 0.000291$

29.   (a) $\frac{\pi}{3} = \frac{\pi}{3}\left(\frac{180}{\pi}\right)°$

$= 60°$

(c) Reference angle is itself: $\frac{\pi}{3} = 60°$

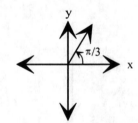

33.   (a) $-\frac{7\pi}{6} = -\frac{7\pi}{6}\left(\frac{180}{\pi}\right)°$

$= -210°$

(c) $-210° = -210° + 360°$

$= 150°$

Reference angle of 150° is $180° - 150° = 30°$

$30° = 30\left(\frac{\pi}{180}\right)$

$= \frac{\pi}{6}$ radians

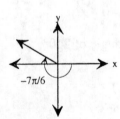

37. (a) $4\pi = 4\pi \left(\frac{180}{\pi}\right)^\circ$

$\qquad = 720^\circ$

(c) $720^\circ = 720^\circ - 360^\circ$

$\qquad = 360^\circ$

$360^\circ = 360^\circ - 360^\circ$

$\qquad = 0^\circ$

Reference angle is 0° or 0 radians

41. $1 = 1\left(\dfrac{180}{\pi}\right)^\circ$

$\qquad = \left(\dfrac{180}{\pi}\right)^\circ$

$\qquad = 57.3^\circ$

45. $0.75 = 0.75\left(\dfrac{180}{\pi}\right)^\circ$

$\qquad = \left(\dfrac{135}{\pi}\right)^\circ$

$\qquad = 43.0^\circ$

49. Since $\dfrac{4\pi}{3}$ terminates in QIII, its sine will be negative.

$\widehat{\theta} = \dfrac{4\pi}{3} - \pi$

$\qquad = \dfrac{\pi}{3}$

$\sin \dfrac{4\pi}{3} = -\sin \dfrac{\pi}{3} \qquad \left(\dfrac{\pi}{3} = 60^\circ\right)$

$\qquad = -\dfrac{\sqrt{3}}{2}$

53. Since $\dfrac{2\pi}{3}$ terminates in QII, its secant will be negative.

$\widehat{\theta} = \pi - \dfrac{2\pi}{3}$

$\qquad = \dfrac{\pi}{3}$

$\sec \dfrac{2\pi}{3} = -\sec \dfrac{\pi}{3}$

$\qquad = -2$

57. Since $-\dfrac{\pi}{4}$ terminates in QIV, its sine will be negative.

$$\theta = -\frac{\pi}{4} + 2\pi$$

$$= \frac{7\pi}{4}$$

$$\widehat{\theta} = 2\pi - \frac{7\pi}{4}$$

$$= \frac{\pi}{4}$$

$$4\sin\left(-\frac{\pi}{4}\right) = -4\sin\frac{\pi}{4}$$

$$= -4\left(\frac{\sqrt{2}}{2}\right)$$

$$= -2\sqrt{2}$$

61. $2\cos\dfrac{\pi}{6} = 2\left(\dfrac{\sqrt{3}}{2}\right)$

$$= \sqrt{3}$$

65. $6\cos 3\left(\dfrac{\pi}{6}\right) = 6\cos\dfrac{\pi}{2}$

$$= 6(0)$$

$$= 0$$

69. $4\cos\left[2\left(\dfrac{\pi}{6}\right) + \dfrac{\pi}{3}\right] = 4\cos\left(\dfrac{\pi}{3} + \dfrac{\pi}{3}\right) = 4\cos\dfrac{2\pi}{3}$

Since $\dfrac{2\pi}{3}$ terminates in QII, its cosine will be negative.

$$\widehat{\theta} = \pi - \frac{2\pi}{3}$$

$$= \frac{\pi}{3}$$

$$4\cos\frac{2\pi}{3} = -4\cos\frac{\pi}{3}$$

$$= -4\left(\frac{1}{2}\right)$$

$$= -2$$

73. For $x = 0$, $\quad y = 2\sin 0$

$$= 2(0)$$

$$= 0 \qquad (x, y) = (0, 0)$$

For $x = \dfrac{\pi}{2}$, $y = 2\sin\dfrac{\pi}{2}$

$= 2(1)$

$= 2$ $(x, y) = (\dfrac{\pi}{2}, 2)$

For $x = \pi$, $y = 2\sin\pi$

$= 2(0)$

$= 0$ $(x, y) = (\pi, 0)$

For $x = \dfrac{3\pi}{2}$, $y = 2\sin\dfrac{3\pi}{2}$

$= 2(-1)$

$= -2$ $(x, y) = (\dfrac{3\pi}{2}, -2)$

For $x = 2\pi$, $y = 2\sin 2\pi$

$= 2(0)$

$= 0$ $(x, y) = (2\pi, 0)$

77. For $x = \dfrac{\pi}{2}$, $y = \sin(\dfrac{\pi}{2} - \dfrac{\pi}{2})$

$= \sin 0$

$= 0$ $(x, y) = (\dfrac{\pi}{2}, 0)$

For $x = \pi$, $y = \sin(\pi - \dfrac{\pi}{2})$

$= \sin\dfrac{\pi}{2}$

$= 1$ $(x, y) = (\pi, 1)$

For $x = \dfrac{3\pi}{2}$, $y = \sin(\dfrac{3\pi}{2} - \dfrac{\pi}{2})$

$= \sin\pi$

$= 0$ $(x, y) = (\dfrac{3\pi}{2}, 0)$

For $x = 2\pi$, $y = \sin(2\pi - \dfrac{\pi}{2})$

$= \sin\dfrac{3\pi}{2}$

$= -1$ $(x, y) = (2\pi, -1)$

For $x = \dfrac{5\pi}{2}$, $y = \sin(\dfrac{5\pi}{2} - \dfrac{\pi}{2})$

$= \sin 2\pi$

$= 0$ $(x, y) = (\dfrac{5\pi}{2}, 0)$

81. $(x, y) = (1, -3)$

$$r = \sqrt{x^2 + y^2} = \sqrt{(1)^2 + (-3)^2} = \sqrt{1 + 9} = \sqrt{10}$$

$$\sin \theta = \frac{y}{r} = \frac{-3}{\sqrt{10}} \qquad \cot \theta = \frac{x}{y} = \frac{1}{-3}$$

$$\cos \theta = \frac{x}{r} = \frac{1}{\sqrt{10}} \qquad \sec \theta = \frac{r}{x} = \frac{\sqrt{10}}{1} = \sqrt{10}$$

$$\tan \theta = \frac{y}{x} = \frac{-3}{1} = -3 \qquad \csc \theta = \frac{r}{y} = \frac{\sqrt{10}}{-3}$$

85. $\sin \theta = \dfrac{1}{2}$ and θ terminates in QII.

$$\cos \theta = -\sqrt{1 - \sin^2 \theta} \qquad \text{because } \cos \theta \text{ is negative in QII}$$

$$= -\sqrt{1 - (\tfrac{1}{2})^2}$$

$$= -\sqrt{1 - \frac{1}{4}}$$

$$= -\sqrt{\frac{3}{4}}$$

$$= -\frac{\sqrt{3}}{2}$$

$$\tan \theta = \frac{\sin \theta}{\cos \theta} \qquad\qquad \cot \theta = \frac{1}{\tan \theta}$$

$$= \frac{1/2}{-\sqrt{3}/2} \qquad\qquad = \frac{1}{-1/\sqrt{3}}$$

$$= -\frac{1}{\sqrt{3}} \qquad\qquad = -\sqrt{3}$$

$$\sec \theta = \frac{1}{\cos \theta} \qquad\qquad \csc \theta = \frac{1}{\sin \theta}$$

$$= \frac{1}{-\sqrt{3}/2} \qquad\qquad = \frac{1}{1/2}$$

$$= -\frac{2}{\sqrt{3}} \qquad\qquad = 2$$

Problem Set 3.3

1. The point on the unit circle is $(-\frac{\sqrt{3}}{2}, \frac{1}{2})$.

$$\sin 150° = \frac{1}{2}$$

$$\cos 150° = \frac{-\sqrt{3}}{2}$$

$$\tan 150° = \frac{\sin 150°}{\cos 150°} = \frac{1/2}{-\sqrt{3}/2} = -\frac{1}{\sqrt{3}}$$

$$\sec 150° = \frac{1}{\cos 150°} = \frac{1}{-\sqrt{3}/2} = -\frac{2}{\sqrt{3}}$$

$$\csc 150° = \frac{1}{\sin 150°} = \frac{1}{1/2} = 2$$

5. The point on the unit circle is $(-1, 0)$.

$$\sin 180° = 0$$

$$\cos 180° = -1$$

$$\tan 180° = \frac{\sin 180°}{\cos 180°} = \frac{0}{-1} = 0$$

$$\cot 180° = \frac{\cos 180°}{\sin 180°} = \frac{-1}{0} \text{ (undefined)}$$

$$\sec 180° = \frac{1}{\cos 180°} = \frac{1}{-1} = -1$$

$$\csc 180° = \frac{1}{\sin 180°} = \frac{1}{0} \text{ (undefined)}$$

9. $\cos(-60°) = \cos 60°$ cosine is an even function from the unit circle

$$= \frac{1}{2}$$

13. $\sin(-30°) = -\sin 30°$ sine is an odd function from the unit circle

$$= -\frac{1}{2}·$$

17. On the unit circle, we locate all points with a y-coordinate of $\frac{1}{2}$. The angles associated with these

points are $\frac{\pi}{6}$ and $\frac{5\pi}{6}$.

21. We look for points on the unit circle where the ratio, $\frac{y}{x}$, equals $-\sqrt{3}$. The angles associated

with these points are $\frac{2\pi}{3}$ and $\frac{5\pi}{3}$.

25. $\sin(-\theta) = -\sin\theta$ sine is an odd function

$$= -(-\frac{1}{3})$$

$$= \frac{1}{3}$$

29. $\sin(2\pi + \frac{\pi}{6}) = \sin\frac{\pi}{6}$

$$= \frac{1}{2}$$

33. $\sin\frac{13\pi}{6} = \sin(2\pi + \frac{\pi}{6})$

$$= \sin\frac{\pi}{6}$$

$$= \frac{1}{2}$$

37. $\tan(-\theta) = \dfrac{\sin(-\theta)}{\cos(-\theta)}$

$$= \frac{-\sin\theta}{\cos\theta}$$

$$= -\tan\theta$$

Therefore, the tangent is an odd function.

41. $\sin(-\theta)\sec(-\theta)\cot(-\theta) = \dfrac{\sin(-\theta)}{1} \cdot \dfrac{1}{\cos(-\theta)} \cdot \dfrac{\cos(-\theta)}{\sin(-\theta)} = 1$

49. $A = 90° - 22° = 68°$

$\tan 22° = \dfrac{320}{a}$	$\sin 22° = \dfrac{320}{c}$
$a = \dfrac{320}{\tan 22°}$	$c = \dfrac{320}{\sin 22°}$
$= \dfrac{320}{0.4040}$	$= \dfrac{320}{0.3746}$
$= 790$	$= 850$

53.

$$b = \sqrt{c^2 - a^2}$$

$$= \sqrt{(6.21)^2 - (4.37)^2}$$

$$= \sqrt{19.4672}$$

$$= 4.41$$

$$\sin A = \frac{4.37}{6.21} \qquad\qquad B = 90° - 44.7°$$

$$= 0.7037 \qquad\qquad\qquad = 45.3°$$

$$A = 44.7°$$

Problem Set 3.4

1. $s = r\,\theta$ Formula for arc length

    $= 3(2)$ Substitute given values

    $= 6$ inches   Simplify

5. $s = r\,\theta$ Formula for arc length

    $= 12\left(\dfrac{\pi}{6}\right)$ Substitute given values

    $= 2\pi$ centimeters   Simplify

    $= 6.28$ centimeters   Rounded to 3 significant digits

9. First, we must change θ to radians by multiplying by $\dfrac{\pi}{180}$.

    $s = r\,\theta$ Formula for arc length

    $= 10\,(240)(\dfrac{\pi}{180})]$ Substitute given values (θ in radians)

    $= \dfrac{40\pi}{3}$ inches   Simplify

    $= 41.9$ inches   Rounded to 3 significant digits

13. First, we find θ: $\dfrac{\theta}{2\pi} = \dfrac{1}{6}$ One complete rotation is 6 hours or 2π radians

              $\theta = \dfrac{2\pi}{6}$ Multiply both sides by 2π

              $= \dfrac{\pi}{3}$ Simplify

Also, the radius is $200 + 4{,}000$ or $4{,}200$ miles.

Therefore, $s = r\theta = 4{,}200(\frac{\pi}{3}) = 1{,}400\pi$ miles $= 4{,}400$ miles

17. First we convert 0.5° to radians by multiplying by $\frac{\pi}{180}$. Then we apply the formula for arc length.

$$s = r\theta$$
$$= 240,000 \left[(0.5)\left(\frac{\pi}{180}\right)\right]$$
$$= \frac{2,000\pi}{3} \text{ miles}$$
$$= 2,100 \text{ miles}$$

21. $\quad s = r\theta$
$$= 125\left[(220)\left(\frac{\pi}{180}\right)\right]$$
$$= \frac{1,375\pi}{9} \text{ ft}$$
$$= 480.0 \text{ ft}$$

25. $\quad r = \dfrac{s}{\theta}$ Formula for arc length

$$= \frac{3}{6} \qquad\qquad \text{Substitute given values}$$
$$= 0.5 \text{ ft} \qquad \text{Simplify}$$

29. $\quad r = \dfrac{s}{\theta}$ Formula for arc length

$$= \frac{\pi}{\pi/4} \qquad\quad \text{Substitute given values}$$
$$= 4 \text{ cm} \qquad\quad \text{Simplify}$$

33. First, we convert θ to radians by multiplying by $\frac{\pi}{180}$.

$$r = \frac{s}{\theta} \qquad\qquad\qquad \text{Formula for arc length}$$
$$= \frac{4}{225\left(\frac{\pi}{180}\right)} \qquad \text{Substitute given values}$$
$$= \frac{16}{5\pi} \text{ or } 1.02 \text{ km} \qquad \text{Simplify}$$

37. $A = \dfrac{1}{2}r^2\theta$ Formula for area of a sector

$= \dfrac{1}{2}(4)^2(2.4)$ Substitute given values

$= 19.2 \text{ in}^2$ Simplify

41. $A = \dfrac{1}{2}r^2\theta$ Formula for area of a sector

$= \dfrac{1}{2}(5)^2(15 \cdot \dfrac{\pi}{180})$ Substitute given values (θ in radians)

$= \dfrac{25\pi}{24}$ or 3.27 m^2 Simplify

45. $A = \dfrac{1}{2}r^2\theta$ Formula for area of a sector

$\dfrac{\pi}{3} = \dfrac{1}{2}r^2(30 \cdot \dfrac{\pi}{180})$ Substitute given values (θ in radians)

$\dfrac{\pi}{3} = \dfrac{\pi}{12}r^2$ Simplify

$4 = r^2$ Solve for r

$r = 2 \text{ cm}$

49. $A = \dfrac{1}{2}r^2\theta$ Formula for area of a sector

$= \dfrac{1}{2}(60)^2 (90 \cdot \dfrac{\pi}{180})$ Substitute given values (θ in radians)

$= 900\pi$ or $2{,}830 \text{ ft}^2$ Simplify

53. $\tan 12° = \dfrac{x}{5.2}$

$x = 5.2 \tan 12°$

$= 5.2(0.2126)$

$= 1.1$

$\tan 13° = \dfrac{y}{5.2}$

$y = 5.2 \tan 13°$

$= 5.2(0.2309)$

$= 1.2$

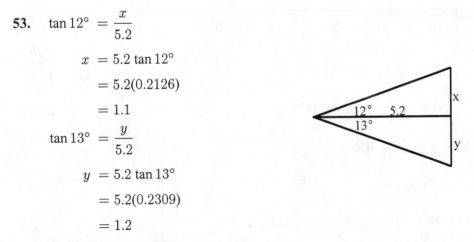

The vertical dimension of the mirror is

$x + y = 1.1 + 1.2 = 2.3 \text{ ft}$

57.

$$\theta = 90° - 43.2° \quad \tan 14.5° = \frac{y}{x}$$

$$= 46.8° \qquad\qquad = \frac{y}{91.6}$$

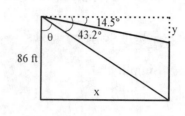

$$\tan\theta = \frac{x}{86}$$

$$x = 86\tan 46.8° \qquad y = 91.6\tan 14.5°$$

$$= 86(1.0649) \qquad\quad = 54(0.2586)$$

$$= 91.6 \qquad\qquad\quad = 23.7$$

The height of the building is $86 - y = 86 - 23.7$

$$= 62.3 \text{ ft}$$

Problem Set 3.5

1. $\quad v = \dfrac{s}{t}$ — Formula for linear velocity

$\quad\quad = \dfrac{3}{2}$ — Substitute given values

$\quad\quad = 1.5$ feet per minute — Simplify

5. $\quad v = \dfrac{s}{t}$ — Formula for linear velocity

$\quad\quad = \dfrac{30}{2}$ — Substitute given values

$\quad\quad = 15$ miles per hour — Simplify

9. $\quad s = vt$ — Formula for linear velocity

$\quad\quad = 45(\dfrac{1}{2})$ — Substitute given values

$\quad\quad = 22.5$ miles — Simplify

13. $\quad \omega = \dfrac{\theta}{t}$ — Formula for angular velocity

$\quad\quad = \dfrac{2\pi/3}{5}$ — Substitute given values

$\quad\quad = \dfrac{2\pi}{15}$ radians per second — Simplify

$\quad\quad = 0.419$ radians per second — Rounded to 3 significant digits

17. $\omega = \dfrac{\theta}{t}$ Formula for angular velocity

$\quad = \dfrac{8\pi}{3\pi}$ Substitute given values

$\quad = \dfrac{8}{3}$ radians per second   Simplify

$\quad = 2.67$ radians per second   Rounded to 3 significant digits

21. We know that $\tan\theta = \dfrac{d}{100}$.

Therefore, $d = 100\tan\theta$.

Next, we must find θ. We know that $\omega = \dfrac{\theta}{t}$. Therefore, $\theta = \omega t$. We also know that

$\omega = \dfrac{2\pi \text{ radians}}{4 \text{ seconds}} = \dfrac{1}{2}\pi$ radians per second.

We have $\theta = \dfrac{1}{2}\pi t$.

Substituting θ, we get: $\quad d = 100\tan\theta$

$= 100\tan\dfrac{1}{2}\pi t$

When $t = \dfrac{1}{2}$, then $d = 100\tan\left[\dfrac{1}{2}(\pi)\left(\dfrac{1}{2}\right)\right]$

$= 100\tan\dfrac{\pi}{4}$

$= 100(1)$

$= 100$ feet

This problem is continued on the next page.

When $t = \dfrac{3}{2}$, $d = 100\tan\left[\dfrac{1}{2}(\pi)\left(\dfrac{3}{2}\right)\right]$

$= 100\tan\dfrac{3\pi}{4}$

$= 100(-1)$

$= -100$ feet

When $t = 1$, $d = 100\tan\left[\dfrac{1}{2}(\pi)(1)\right]$

$= 100\tan\dfrac{\pi}{2}$

d is undefined because $\tan\dfrac{\pi}{2}$ is undefined.

When $t = 1$, $\theta = \dfrac{\pi}{2}$ and the light rays are parallel to the wall.

---

25. First, we find θ, using the formula for angular velocity. Then we apply the formula for arc length.

$$\theta = \omega t \qquad \text{Formula for angular velocity}$$

$$= \frac{3\pi}{2}(30) \qquad \text{Substitute given values}$$

$$= 45\pi \qquad \text{Simplify}$$

$$s = r\theta \qquad \text{Formula for arc length}$$

$$= 4(45\pi) \qquad \text{Substitute given values}$$

$$= 180\pi \text{ meters} \quad \text{Simplify}$$

$$= 565 \text{ meters} \quad \text{Rounded to 3 significant digits}$$

In Problems 29 and 33 we convert revolutions per minute to radians per minute by multiplying by 2π.

29. $\omega = 10(2\pi)$ **33.** $\omega = 5.8(2\pi)$

       $= 20\pi$ radians per minute $= 11.6\pi$ radians per minute

       $= 62.8$ radians per minute $= 36.4$ radians per minute

37. Using the relationship between angular velocity and linear velocity, we have

$$\omega = \frac{v}{r}$$

$$= \frac{3}{6}$$

$$= 0.5 \text{ radians per second}$$

41. Angular velocity, ω, is $\frac{1}{24}$ revolutions per hour. To convert this to radians per hour, we multiply by 2π.

$$\omega = \frac{1}{24}(2\pi)$$

$$= \frac{\pi}{12} \text{ radians per hour}$$

$$= 0.262 \text{ radians per hour}$$

45. Using the relationship between angular velocity and linear velocity, we have

$$\omega = \frac{v}{r}$$

$$= \frac{1,100 \text{ ft/sec}}{1 \text{ ft}}$$

$$= 1,100 \text{ radians per second}$$

Then we convert this to revolutions per second by dividing by 2π.

$$\omega = \frac{1,100}{2\pi}$$

$$= \frac{550}{\pi} \text{ revolutions per second}$$

To convert this to revolutions per minute, we multiply by 60.

$$\omega = \frac{550}{\pi}(60)$$

$$= \frac{33,000}{\pi} \text{ rpm}$$

$$= 10,500 \text{ rpm}$$

49. Angular velocity, ω, is $\frac{1}{15}$ revolutions per minute. To convert to radians per minute, we multiply by 2π.

$$\omega = \frac{1}{15}(2\pi)$$

$$= \frac{2\pi}{15} \text{ radians per minute}$$

$v = r\omega$ Relationship between angular and linear velocities

$$= (\frac{197}{2})(\frac{2\pi}{15}) \qquad \text{Substitute given values}$$

$$= \frac{197\pi}{15} \text{ ft/min}$$

To convert to miles, we divide by 5,280 ft and to convert to hours, we multiply by 60 min.

$$\frac{197\pi}{15} \cdot \frac{1}{5,280} \cdot \frac{60}{1} = 0.47 \frac{\text{mi}}{\text{hr}}$$

53. $v = \frac{s}{t}$ Formula for linear velocity

$$= \frac{16 \text{ km}}{1 \text{ hr}} \qquad \text{Substitute given values}$$

$= 16 \text{ km per hour}$ Simplify

$= 16,000 \text{ meters per hour}$ Change to meters per hour by multiplying by 1,000

$\omega = \frac{v}{r}$ Relationship between angular velocity and linear velocity

$$= \frac{16,000 \text{ m/hr}}{0.3 \text{ m}} \qquad \text{Substitute given values and change radius to meters}$$

$= 53,300 \text{ radians per hour}$ Simplify

57. $|\vec{V}| = \sqrt{(45.5)^2 + (176)^2}$ Pythagorean theorem

$\qquad = \sqrt{2{,}070.25 + 30{,}976}$ Simplify

$\qquad = \sqrt{33{,}046.25}$ Simplify

$\qquad = 182$ mph   Rounded to 3 significant digits

$\tan\theta = \dfrac{45.5}{176}$ Tangent ratio

$\qquad = 0.2585$ Simplify

$\theta = 14.5°$ Rounded to the nearest tenth of a degree

The ground speed is 182 mph at 54.5° from due north.

61. $\theta = 180° + 32.7° = 212.7°$

$|\vec{V}_x| = |\vec{V}|\cos\theta$

$\qquad = 85.5\cos 212.7°$

$\qquad = 85.5(-0.8415)$

$\qquad = -71.9$

$|\vec{V}_y| = |\vec{V}|\sin\theta$

$\qquad = 85.5\sin 212.7°$

$\qquad = 85.5\,(-0.5402)$

$\qquad = -46.2$

The ship has sailed 71.9 miles west and 46.2 miles south.

Chapter 3 Test

1. $\widehat{\theta} = 235° - 180°$
     $\quad = 55°$

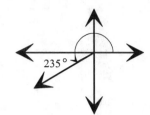

2. $\widehat{\theta} = 180° - 117.8°$
     $\quad = 62.2°$

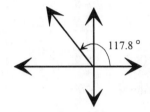

3. $\widehat{\theta} = 410° \, 20' - 360°$
     $\quad = 50° \, 20'$

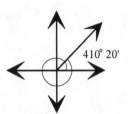

4. $\theta = -225° + 360°$
     $\quad = 135°$
     $\widehat{\theta} = 180° - 135°$
     $\quad = 45°$

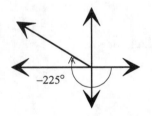

5. $\cot 320° = \dfrac{1}{\tan 320°}$ Reciprocal identity

           $= \dfrac{1}{-0.8391}$ Simplify using a calculator

           $= -1.1918$

6. $\cot(-25°) = \dfrac{1}{\tan(-25°)}$ Reciprocal identity

           $= \dfrac{1}{-\tan 25°}$ Tangent is an odd function

           $= -\dfrac{1}{0.4663}$ Simplify, using a calculator

           $= -2.1445$

7. $\csc(-236.7°) = \dfrac{1}{\sin(-236.7°)}$ Reciprocal identity

$\qquad\qquad = \dfrac{1}{-\sin 236.7°}$ Sine is an odd function

$\qquad\qquad = \dfrac{1}{0.8358}$ Simplify, using a calculator

$\qquad\qquad = 1.1964$

8. $\sec 322.3° = \dfrac{1}{\cos 322.3°}$ Reciprocal identity

$\qquad\qquad = \dfrac{1}{0.7912}$ Simplify, using a calculator

$\qquad\qquad = 1.2639$

9. $140°\ 20' = 140° + \left(\dfrac{20}{60}\right)°$

$\qquad\qquad = 140.33°$

$\sec 140.33° = \dfrac{1}{\cos 140.33°}$ Reciprocal identity

$\qquad\qquad = \dfrac{1}{-0.7698}$ Simplify using a calculator

$\qquad\qquad = -1.2991$

10. $188°\ 50' = 188° + \left(\dfrac{50}{60}\right)°$

$\qquad\qquad = 188.833°$

$\csc 188.833° = \dfrac{1}{\sin 188.833°}$ Reciprocal identity

$\qquad\qquad = \dfrac{1}{-0.1536}$ Simplify, using a calculator

$\qquad\qquad = -6.5121$

11. $\sin\widehat{\theta} = 0.1045$ $\theta = 180° - 6°$

$\qquad \widehat{\theta} = 6°$ $= 174°$

12. $\cos\widehat{\theta} = 0.4772$ $\theta = 180° + 61.5°$

   $\widehat{\theta} = 61.5°$ $= 241.5°$

13. $\cot\widehat{\theta} = 0.9659$

   $\tan\widehat{\theta} = \dfrac{1}{0.9659}$

   Scientific Calculator: 0.9659 | 1/x | | inv | | tan |

   Graphing Calculator: | 2nd | | tan | | (| 1 | ÷ | 0.9659 |) | | ENTER |

   Answer: 46.0°

   Since θ is in QIII, $\theta = 180° + \widehat{\theta}$

   $= 180° + 46.0°$

   $= 226°$

14. $\sec\widehat{\theta} = 1.545$

   $\cos\widehat{\theta} = \dfrac{1}{1.545}$

   Scientific Calculator: 1.545 | 1/x | | inv | | cos |

   Graphing Calculator: | 2nd | | cos | | (| 1 | ÷ | 1.545 |) | | ENTER |

   Answer: 49.7°

   Since θ is in QIV, $\theta = 360° - \widehat{\theta}$

   $= 360° - 49.7°$

   $= 310.3°$

15. $\widehat{\theta} = 225° - 180°$

   $= 45°$

   $\sin 225° = -\sin 45°$ θ is in QIII

   $= -\dfrac{1}{\sqrt{2}}$ or $-\dfrac{\sqrt{2}}{2}$

16. $\widehat{\theta} = 180° - 135°$

   $= 45°$

   $\cos 135° = -\cos 45°$ θ is in QII

   $= -\dfrac{1}{\sqrt{2}}$ or $-\dfrac{\sqrt{2}}{2}$

17. $\hat{\theta} = 360° - 330°$

$\qquad = 30°$

$\tan 330° = -\tan 30°$ $\qquad \theta$ is in QIV

$\qquad\qquad = -\dfrac{1}{\sqrt{3}}$

18. $\hat{\theta} = 390° - 360°$

$\qquad = 30°$

$\sec 390° = \sec 30°$ $\qquad \theta$ is in QI

$\qquad\qquad = \dfrac{1}{\cos 30°}$

$\qquad\qquad = \dfrac{1}{\sqrt{3}/2}$

$\qquad\qquad = \dfrac{2}{\sqrt{3}}$

19. $\theta = 250 \cdot \dfrac{\pi}{180} = \dfrac{25\pi}{18}$ radians

20. $\theta = -390 \cdot \dfrac{\pi}{180} = -\dfrac{13\pi}{6}$ radians

21. $\theta = \dfrac{4\pi}{3}\left(\dfrac{180}{\pi}\right)°$

$\qquad = 240°$

22. $\theta = \dfrac{7\pi}{12}\left(\dfrac{180}{\pi}\right)$

$\qquad = 105°$

23. $\hat{\theta} = \pi - \dfrac{2\pi}{3}$

$\qquad = \dfrac{\pi}{3}$

$\sin \dfrac{2\pi}{3} = \sin \dfrac{\pi}{3}$ $\qquad \theta$ is in QII

$\qquad\qquad = \dfrac{\sqrt{3}}{2}$

24. $\hat{\theta} = \pi - \dfrac{2\pi}{3} = \dfrac{\pi}{3}$

$\cos \dfrac{2\pi}{3} = -\cos \dfrac{\pi}{3}$ θ is in QII

$= -\dfrac{1}{2}$

25. $4 \cos\left(-\dfrac{3\pi}{4}\right) = 4 \cos \dfrac{3\pi}{4}$ Cosine is an even function

$= 4\left(-\cos \dfrac{\pi}{4}\right)$ θ is in QII

$= 4\left(-\dfrac{1}{\sqrt{2}}\right)$ Substitute exact value

$= -\dfrac{4}{\sqrt{2}}$ Simplify

$= -2\sqrt{2}$ Rationalize the denominator

26. $2 \cos\left(-\dfrac{5\pi}{3}\right) = 2 \cos \dfrac{5\pi}{3}$ Cosine is an even function

$= 2 \cos \dfrac{\pi}{3}$ θ is in QIV

$= 2\left(\dfrac{1}{2}\right)$ Substitute exact value

$= 1$ Simplify

27. $\sec \dfrac{5\pi}{6} = -\sec \dfrac{\pi}{6}$ θ is in QII

$= -\dfrac{2}{\sqrt{3}}$ Substitute exact value

28. $\csc \dfrac{5\pi}{6} = \csc \dfrac{\pi}{6}$ θ is in QII

$= 2$ Substitute exact value

29. $2 \cos\left(3x - \dfrac{\pi}{2}\right) = 2 \cos\left[3\left(\dfrac{\pi}{3}\right) - \dfrac{\pi}{2}\right]$

$= 2 \cos\left(\pi - \dfrac{\pi}{2}\right)$

$= 2 \cos \dfrac{\pi}{2}$

$= 2(0)$

$= 0$

30. $4\sin(2x + \frac{\pi}{4}) = 4\sin[2(\frac{\pi}{4}) + \frac{\pi}{4}]$

$$= 4\sin\frac{3\pi}{4}$$

$$= 4\sin\frac{\pi}{4} \qquad \theta \text{ is in QII}$$

$$= 4(\frac{\sqrt{2}}{2})$$

$$= 2\sqrt{2}$$

31. $\cot(-x) = \dfrac{\cos(-x)}{\sin(-x)}$ Ratio identity

$$= \frac{\cos x}{-\sin x} \qquad \text{Cosine is an even function and sine is an odd function}$$

$$= -\frac{\cos x}{\sin x}$$

$$= -\cot x$$

Therefore, cotangent is an odd function.

32. $\sin(-\theta)\sec(-\theta)\cot(-\theta) = \sin(-\theta) \cdot \dfrac{1}{\cos(-\theta)} \cdot \dfrac{\cos(-\theta)}{\sin(-\theta)} = 1$

33. $s = r\theta$

$$= 12(\frac{\pi}{6})$$

$$= 2\pi \text{ meters}$$

$$= 6.28 \text{ meters}$$

34. $s = r\theta$

$$= 6[60(\frac{\pi}{180})] \qquad \theta \text{ must be in radians}$$

$$= 6(\frac{\pi}{3})$$

$$= 2\pi \text{ or } 6.28 \text{ ft}$$

35. $r = \dfrac{s}{\theta}$

$$= \frac{\pi}{\pi/4}$$

$$= 4 \text{ cm}$$

36. $r = \dfrac{s}{\theta}$

$\quad = \dfrac{\pi/4}{2\pi/3}$

$\quad = \dfrac{3}{8}$ or 0.375 cm

37. $A = \dfrac{1}{2} r^2 \theta$ where $\theta = 90(\dfrac{\pi}{180}) = \dfrac{\pi}{2}$ radians

$\quad = \dfrac{1}{2}(4)^2(\dfrac{\pi}{2})^2$

$\quad = 4\pi$ in^2

$\quad = 12.6$ in^2

38. $A = \dfrac{1}{2} r^2 \theta$

$\quad = \dfrac{1}{2}(3)^2(2.4)$

$\quad = 10.8$ cm^2

39. In 30 minutes, θ is π radians.

$s = r\theta$

$\quad = 2\pi$ or 6.28 cm

40. $r = \dfrac{s}{\theta} \qquad\qquad A = \dfrac{1}{2} r^2 \theta$

$\quad = \dfrac{8}{4} \qquad\qquad\quad = \dfrac{1}{2}(2)^2(4)$

$\quad = 2 \qquad\qquad\quad\; = 8$ in^2

41. $s = vt$

$\quad = 30(3)$

$\quad = 90$ feet

42. $s = vt$

$\quad = 66(60) \qquad$ Time must be in seconds

$\quad = 3960$ ft

---

43. $\theta = \omega t$ — Formula for angular velocity

$\qquad = 4(6)$ — Substitute given values

$\qquad = 24$ radians — Simplify

$s = r\theta$ — Formula for arc length

$\qquad = 3(24)$ — Substitute given values

$\qquad = 72$ in — Simplify

44. $\theta = \omega t$ — Formula for angular velocity

$\qquad = \dfrac{3\pi}{4}(20)$ — Substitute given values

$\qquad = 15\pi$ — Simplify

$s = r\theta$ — Formula for arc length

$\qquad = 8(15\pi)$ — Substitute given values

$\qquad = 120\pi$ or 377 ft — Simplify

45. $\omega = 6(2\pi)$

$\qquad = 12\pi$ rad/min

$\qquad = 37.7$ rad/min

46. $\omega = 2(2\pi)$

$\qquad = 4\pi$ rad/min

$\qquad = 12.6$ rad/min

47. $\omega = \dfrac{v}{r}$ — Relationship between angular and linear velocities

$\qquad = \dfrac{5}{10}$ — Substitute given values

$\qquad = 0.5$ rad/sec — Simplify

48. $\omega = \dfrac{v}{r}$ — Relationship between angular and linear velocities

$\qquad = \dfrac{5}{3}$ rad/sec — Substitute given values

49. $\omega = 20(2\pi)$

$= 40\pi$ rad/min

$v = r\omega$

$= 2(40\pi)$

$= 80\pi$ ft/min

$= 251$ ft/min

50. $\omega = 10(2\pi)$

$= 20\pi$ rad/min

$v = r\omega$

$= 1(20\pi)$

$= 20\pi$ ft/min

$= 62.8$ ft/min

51. $\omega_1 = \dfrac{v}{r_1}$ $\qquad\qquad$ $\omega_2 = \dfrac{v}{r_2}$

$\quad = \dfrac{24}{6}$ $\qquad\qquad\quad = \dfrac{24}{8}$

$\quad = 4$ rad/sec (for 6 cm pulley) $\quad = 3$ rad/sec (for 8 cm pulley)

52. $\omega = 900(2\pi)$

$= 1,800\pi$ rad/min

$v = r\,\omega$

$= 1.5(1800\,\pi)$

$= 2,700\pi$ ft/min

$= 8,480$ ft/min

CHAPTER 4 Graphing and Inverse Functions

Problem Set 4.1

9. Refer to the graph in Problem 3. Find the x-value corresponding to $y = -1$.

13. Refer to the graph in Problem 5. Find the x-values corresponding to $y = 1$.

25. Refer to the graph in Problem 19. Find the x-values corresponding to $y = -1$.

29. Refer to the graph in Problem 21. Find the x-values corresponding to $y = 1$.

33. Refer to the graphs in Problems 1 and 18. Find the x-values corresponding to $y = 0$. You will notice the pattern that exists: $x = \dfrac{\pi}{2} + k\pi$ where k is any integer.

37. Refer to the graphs in Problems 5 and 22. Find the x-values corresponding to $y = 0$. You will notice the pattern that exists: $x = k\pi$ where k is any integer.

41. The amplitude is 2 because the greatest y-value is 2 and the least y-value is -2. The period is 2 because the graph repeats itself every 2 units, or $f(x + 2) = f(x)$.

49.
$$\cos\theta\,\tan\theta = \cos\theta \cdot \frac{\sin\theta}{\cos\theta} \qquad \text{Ratio identity}$$

$$= \frac{\cos\theta\,\sin\theta}{\cos\theta} \qquad \text{Multiplication of fractions}$$

$$= \sin\theta \qquad \text{Division of common factor}$$

53.
$$\csc\theta + \sin(-\theta) = \csc\theta - \sin\theta \qquad \text{Sine is an odd function}$$

$$= \frac{1}{\sin\theta} - \sin\theta \qquad \text{Reciprocal identity}$$

$$= \frac{1}{\sin\theta} - \sin\theta \cdot \frac{\boldsymbol{\sin\theta}}{\boldsymbol{\sin\theta}} \qquad \text{LCD is } \sin\theta$$

$$= \frac{1}{\sin\theta} - \frac{\sin^2\theta}{\sin\theta} \qquad \text{Multiplication of fractions}$$

$$= \frac{1 - \sin^2 \theta}{\sin \theta} \qquad \text{Subtraction of fractions}$$

$$= \frac{\cos^2 \theta}{\sin \theta} \qquad \text{Pythagorean identity}$$

57. $\dfrac{\pi}{4} = \dfrac{\pi}{4}(\dfrac{180}{\pi})°$

$\qquad = 45°$

61. $\dfrac{11\pi}{6} = \dfrac{11\pi}{6}(\dfrac{180}{\pi})°$

$\qquad = 330°$

Problem Set 4.2

1. $y = 6 \sin x$ Amplitude $= 6$

$\qquad$ Period $= \dfrac{2\pi}{1} = 2\pi$

5. $y = \cos \dfrac{1}{3} x$ Amplitude $= 1$

$\qquad$ Period $= \dfrac{2\pi}{1/3} = 6\pi$

9. $y = \sin \pi x$ Amplitude $= 1$

$\qquad$ Period $= \dfrac{2\pi}{\pi} = 2$

13. $y = 4 \sin 2x$ Amplitude $= 4$

$\qquad$ Period $= \dfrac{2\pi}{2} = \pi$

17. $y = 3 \sin \dfrac{1}{2} x$ Amplitude $= 3$

$\qquad$ Period $= \dfrac{2\pi}{1/2} = 4\pi$

21. $y = \dfrac{1}{2} \sin \dfrac{\pi}{2} x$ Amplitude $= \dfrac{1}{2}$

$\qquad$ Period $= \dfrac{2\pi}{\pi/2} = 4$

25. $y = -3 + 2 \cos 4x$ Amplitude $= 2$

$\qquad$ Period $= \dfrac{2\pi}{4} = \dfrac{\pi}{2}$

$\qquad$ Vertical translation $= -3$

The graph of $y = -3 + 2 \cos 4x$ is the graph of $y = 2 \cos 4x$ with all points moved down 3 units.

29. $y = -1 + \dfrac{1}{2} \cos 3x$ Amplitude $= \dfrac{1}{2}$

$\qquad$ Period $= \dfrac{2\pi}{3}$

$\qquad$ Vertical translation $= -1$

The graph of $y = -1 + \dfrac{1}{2} \cos 3x$ is the graph of $y = \dfrac{1}{2} \cos 3x$ with all points moved down 1 unit.

33. $y = 2 \sin \pi x$ $\quad$ Amplitude $= 2$

$\quad\quad\quad\quad\quad\quad\quad\quad$ Period $= \dfrac{2\pi}{\pi} = 2$

37. $y = -3 \cos \dfrac{1}{2} x$ $\quad$ Amplitude $= |-3| = 3$

$\quad\quad\quad\quad\quad\quad\quad\quad$ Period $= \dfrac{2\pi}{1/2} = 4\pi$

41. $I = 20 \sin 120\pi t$ $\quad$ Amplitude $= 20$

$\quad\quad\quad\quad\quad\quad\quad\quad$ Period $= \dfrac{2\pi}{120\pi} = \dfrac{1}{60}$

The maximum value of I is 20 amperes. One complete cycle takes $\frac{1}{60}$ seconds.

45. $y = 2 \csc 3x$ $\quad$ Range: $y \geq 2$ and $y \leq -2$

$\quad\quad\quad\quad\quad\quad\quad\quad$ Period $= \dfrac{2\pi}{3}$

Since $y = 2 \csc 3x$ is the reciprocal of $y = 2 \sin 3x$, we sketch $y = 2 \sin 3x$ using a dotted line. We draw a vertical asymptote wherever $y = \sin 3x$ is 0. Using the graph of $y = 2 \sin 3x$ and the vertical asymptotes, we sketch its reciprocal.

49. $y = 3 \sec \dfrac{1}{2} x$ $\quad$ Range: $y \geq 3$ and $y \leq -3$

$\quad\quad\quad\quad\quad\quad\quad\quad$ Period $= \dfrac{2\pi}{1/2} = 4\pi$

Since $y = 3 \sec \dfrac{1}{2} x$ is the reciprocal of $y = 3 \cos \dfrac{1}{2} x$, we sketch $y = 3 \cos \dfrac{1}{2} x$ using a dotted line. We draw a vertical asymptote wherever $y = 3 \cos \dfrac{1}{2} x$ is 0. Using the graph of $y = 3 \cos \dfrac{1}{2} x$ and the vertical asymptotes, we sketch its reciprocal.

53. $d = 100 \tan \dfrac{\pi}{2} t$

t	0	$\frac{1}{2}$	1	$\frac{3}{2}$	2	$\frac{5}{2}$	3	$\frac{7}{2}$	4
d	0	100	N.D.	-100	0	100	N.D.	-100	0

57. $y = \cot 2x$ Period $= \dfrac{\pi}{2}$

The graph will go through 2 complete cycles every π units.

61. $l = 10 \sec \pi t$

t	0	$\frac{1}{4}$	$\frac{1}{2}$	$\frac{3}{4}$	1	$\frac{5}{4}$	$\frac{3}{2}$	$\frac{7}{4}$	2
l	10	$10\sqrt{2}$	N.D.	$-10\sqrt{2}$	-10	$-10\sqrt{2}$	N.D.	$10\sqrt{2}$	10

If $t = 0$, $l = 10 \sec 0$
$$= 10(1)$$
$$= 10$$

If $t = 1$, $l = 10 \sec \pi$
$$= 10(-1)$$
$$= -10$$

If $t = \dfrac{1}{4}$, $l = 10 \sec \dfrac{\pi}{4}$
$$= 10\sqrt{2}$$
$$= 14.14$$

If $t = \dfrac{5}{4}$, $l = 10 \sec \dfrac{5\pi}{4}$
$$= 10\,(-\sqrt{2})$$
$$= -14.14$$

If $t = \dfrac{1}{2}$, $l = 10 \sec \dfrac{\pi}{2}$
l is undefined

If $t = \dfrac{3}{2}$, $l = 10 \sec \dfrac{3\pi}{2}$
l is undefined

If $t = \dfrac{3}{4}$, $l = 10 \sec \dfrac{3\pi}{4}$
$$= 10(-\sqrt{2})$$
$$= -14.14$$

If $t = \dfrac{7}{4}$, $l = 10 \sec \dfrac{7\pi}{4}$
$$= 10\sqrt{2}$$
$$= 14.14$$

If $t = 2$, $l = 10 \sec 2\pi$
$$= 10(1)$$
$$= 10$$

65. $\cos\left(y - \dfrac{\pi}{6}\right) = \cos\left(\dfrac{\pi}{6} - \dfrac{\pi}{6}\right)$
$$= \cos 0$$
$$= 1$$

69.
$$\sin x + \sin y = \sin \frac{\pi}{2} + \sin \frac{\pi}{6}$$
$$= 1 + \frac{1}{2}$$
$$= \frac{3}{2}$$

73. $60° = 60(\frac{\pi}{180})$
$$= \frac{\pi}{3}$$

77. $225° = 225(\frac{\pi}{180})$
$$= \frac{5\pi}{4}$$

Problem Set 4.3

13. Amplitude $= 1$

Period $= \dfrac{2\pi}{\pi} = 2$

Phase Shift $= \dfrac{-\pi/2}{\pi} = -\dfrac{1}{2}$

$A =$ Starting point $= -\dfrac{1}{2}$

$E =$ Ending point $= -\dfrac{1}{2} + 2 = \dfrac{3}{2}$

$C =$ Center point $= \dfrac{1}{2}(-\dfrac{1}{2} + \dfrac{3}{2}) = \dfrac{1}{2}(1) = \dfrac{1}{2}$

$B =$ Average of starting point and center $= \dfrac{1}{2}(-\dfrac{1}{2} + \dfrac{1}{2}) = \dfrac{1}{2}(0) = 0$

$D =$ Average of center and ending point $= \dfrac{1}{2}(\dfrac{1}{2} + \dfrac{3}{2}) = \dfrac{1}{2}(2) = 1$

The 5 points we use on the x-axis are $-\dfrac{1}{2}, 0, \dfrac{1}{2}, 1, \dfrac{3}{2}$.

The 2 points we use on the y-axis are 1 and -1.

17. Amplitude $= 2$

Period $= \dfrac{2\pi}{1/2} = 4\pi$

Phase Shift $= \dfrac{-\pi/2}{1/2} = -\pi$

$A = -\pi$

$E = -\pi + 4\pi = 3\pi$

$$C = \frac{1}{2}(-\pi + 3\pi) = \frac{1}{2}(2\pi) = \pi$$

$$B = \frac{1}{2}(-\pi + \pi) = \frac{1}{2}(0) = 0$$

$$D = \frac{1}{2}(\pi + 3\pi) = \frac{1}{2}(4\pi) = 2\pi$$

The 5 points we use on the x-axis are $-\pi, 0, \pi, 2\pi, 3\pi$.

The 2 points we use on the y-axis are 2 and -2.

21. Amplitude $= 3$

   Period $= \dfrac{2\pi}{\pi/3} = 6$

   Phase Shift $= \dfrac{\pi/3}{\pi/3} = 1$

   $A = 1$

   $E = 1 + 6 = 7$

   $C = \dfrac{1}{2}(1 + 7) = \dfrac{1}{2}(8) = 4$

   $B = \dfrac{1}{2}(1 + 4) = \dfrac{1}{2}(5) = \dfrac{5}{2}$

   $D = \dfrac{1}{2}(4 + 7) = \dfrac{1}{2}(11) = \dfrac{11}{2}$

   The 5 points we use on the x-axis are $1, \dfrac{5}{2}, 4, \dfrac{11}{2}, 7$.

   The 2 points we use on the y-axis are 3 and -3.

25. The graph of $y = -3 + \sin(\pi x + \dfrac{\pi}{2})$ is the graph of $y = \sin(\pi x + \dfrac{\pi}{2})$ (from Problem 13) with all points moved down 3 units.

29. The graph of $y = -2 + 2\sin(\dfrac{1}{2}x + \dfrac{\pi}{2})$ is the graph of $y = 2\sin(\dfrac{1}{2}x + \dfrac{\pi}{2})$ (from Problem 17) with all points moved down 2 units.

33. Amplitude $= 4$

   Period $= \dfrac{2\pi}{2} = \pi$

   Phase Shift $= \dfrac{\pi/2}{2} = \dfrac{\pi}{4}$

   Continued on next page.

$A = \frac{\pi}{4}$

$E = \frac{\pi}{4} + \pi = \frac{5\pi}{4}$

$C = \frac{1}{2}(\frac{\pi}{4} + \frac{5\pi}{4}) = \frac{1}{2}(\frac{3\pi}{2}) = \frac{3\pi}{4}$

$B = \frac{1}{2}(\frac{\pi}{4} + \frac{3\pi}{4}) = \frac{1}{2}(\pi) = \frac{\pi}{2}$

$D = \frac{1}{2}(\frac{3\pi}{4} + \frac{5\pi}{4}) = \frac{1}{2}(2\pi) = \pi$

For one complete cycle, the points we use on the x-axis are $\frac{\pi}{4}, \frac{\pi}{2}, \frac{3\pi}{4}, \pi, \frac{5\pi}{4}$.

The points we use on the y-axis are 4 and -4.

We must extend our graph from $-\frac{\pi}{4}$ to $\frac{3\pi}{2}$.

37. Amplitude $= \frac{2}{3}$

Period $= \frac{2\pi}{3}$

Phase Shift $= \frac{-\pi/2}{3} = \frac{\pi}{6}$

$A = -\frac{\pi}{6}$

$E = -\frac{\pi}{6} + \frac{2\pi}{3} = \frac{\pi}{2}$

$C = \frac{1}{2}(-\frac{\pi}{6} + \frac{\pi}{2}) = \frac{1}{2}(\frac{\pi}{3}) = \frac{\pi}{6}$

$B = \frac{1}{2}(-\frac{\pi}{6} + \frac{\pi}{6}) = \frac{1}{2}(0) = 0$

$D = \frac{1}{2}(\frac{\pi}{6} + \frac{\pi}{2}) = \frac{1}{2}(\frac{2\pi}{3}) = \frac{\pi}{3}$

For one complete cycle, the points we use on the x-axis are $-\frac{\pi}{6}, 0, \frac{\pi}{6}, \frac{\pi}{3}, \frac{\pi}{2}$.

The points we use on the y-axis are $\frac{2}{3}$ and $-\frac{2}{3}$.

We must extend our graph from $-\pi$ to π.

41. First, we will sketch the reciprocal function, $y = \sin(x + \frac{\pi}{4})$ (from Problem 1).

The 5 points on the x-axis are $-\frac{\pi}{4}, \frac{\pi}{4}, \frac{3\pi}{4}, \frac{5\pi}{4}, \frac{7\pi}{4}$. The 2 points on the y-axis are 1 and -1.

We sketch this sine curve using a dotted line. Then we use the sine curve to graph

$y = \csc(x + \frac{\pi}{4})$: the asymptotes will occur where $y = 0$, that is at $-\frac{\pi}{4}, \frac{3\pi}{4}$, and $\frac{7\pi}{4}$.

Using the asymptotes and the sine curve, we sketch the graph.

45. First, we will sketch the reciprocal function, $y = 3\sin(2x + \frac{\pi}{3})$.

Period $= \dfrac{2\pi}{2} = \pi$

Amplitude $= 3$

Phase shift $= -\dfrac{\pi/3}{2} = -\dfrac{\pi}{6}$

The 5 points we use on the x-axis are $-\dfrac{\pi}{6}, \dfrac{\pi}{12}, \dfrac{\pi}{3}, \dfrac{7\pi}{12}, \dfrac{5\pi}{6}$.

The 2 points we use on the y-axis are 3 and -3.

We sketch this sine curve using a dotted line. Then we use the sine curve to graph $y = 3\csc(2x + \pi/3)$: the asymptotes will occur where $y = 0$, that is at $-\pi/6$, $\pi/3$, and $5\pi/6$. Using the asymptotes and the sine curve, we sketch the graph.

49. Period $= \pi$

Phase shift $= \dfrac{\pi}{4}$

Asymptotes will occur at $(0 + \frac{\pi}{4}) + k\pi$ or at $\dfrac{\pi}{4} + k\pi$, where k is an integer.

On the interval from 0 to $\dfrac{5\pi}{4}$, the asymptotes will be at $\dfrac{\pi}{4}$ and $\dfrac{5\pi}{4}$ and the x-intercept will be at $\dfrac{3\pi}{4}$.

53. $s = r\,\theta$

$= 10(\dfrac{\pi}{6})$

$= \dfrac{5\pi}{3}$ cm

57. $r = \dfrac{s}{\theta}$

$= \dfrac{4}{6}$

$= \dfrac{2}{3}$ ft or 8 in

Problem Set 4.4

1. Since the line crosses the y-axis at 1, we know that $b = 1$. The ratio of vertical change to horizontal change between any 2 points is $\dfrac{1}{2}$. Therefore, $m = \dfrac{1}{2}$. The equation of the line must be $y = \dfrac{1}{2}x + 1$.

5. The graph is a sine curve with an amplitude of 1, period 2π, and no phase shift. The equation is $y = \sin x$.

9. The graph is a cosine curve that has been reflected about the x-axis. The amplitude is 3, the period 2π, and no phase shift. The equation is $y = -3\cos x$.

13. The graph is a sine curve with an amplitude of 1, period 6π and no phase shift.

To find B, we set 6π equal to $\dfrac{2\pi}{B}$:

$$6\pi = \frac{2\pi}{B}$$

$$6\pi B = 2\pi$$

$$B = \frac{1}{3}$$

Therefore, the equation is $y = \sin \dfrac{1}{3}\, x$.

17. The graph is a sine curve with an amplitude of 4, period 2, and no phase shift.

To find B, we use $\qquad 2 = \dfrac{2\pi}{B}$

$$2B = 2\pi$$

$$B = \pi$$

Therefore, the equation is $y = 4\sin \pi x$

21. The graph is a sine curve that has been reflected about the x-axis and also moved up 2 units. The amplitude is 4, the period 2, and no phase shift.

To find B, we use $\qquad 2 = \dfrac{2\pi}{B}$

$$2B = 2\pi$$

$$B = \pi$$

Therefore, the equation is $y = 2 - 4\sin \pi x$.

25. The graph is a cosine curve that has been reflected about the x-axis. The amplitude is 2. One complete cycle goes from $-\pi/6$ to $\pi/2$, therefore, the period is $\pi/2 - (-\pi/6) = \dfrac{4\pi}{6} = \dfrac{2\pi}{3}$. The phase shift is $-\pi/6$.

To find B, we use $\dfrac{2\pi}{3} = \dfrac{2\pi}{B}$

$$B = 3$$

To find C, we use $-\dfrac{\pi}{6} = -\dfrac{C}{B}$

$$-\dfrac{\pi}{6} = -\dfrac{C}{3}$$

$$\dfrac{\pi}{2} = C$$

Therefore, the equation is $y = -2\cos\left(3x + \dfrac{\pi}{2}\right)$

29. The graph is the cosine curve from Problem 25 moved up 2 units.

Therefore, the equation is $y = 2 - 2\cos\left(3x + \dfrac{\pi}{2}\right)$

33. $\widehat{\theta} = 360° - 321°$

$= 39°$

37. $\theta = -276° + 360°$

$= 84°$

$\widehat{\theta} = 84°$

41. $\csc\widehat{\theta} = 2.3228$ with θ in QIII

$\sin\widehat{\theta} = \dfrac{1}{2.3228}$

$= 0.4305$

$\widehat{\theta} = 25.5°$

Therefore, $\theta = 180° + 25.5° = 205.5°$

Problem Set 4.5

1. We let $y_1 = 1$ and $y_2 = \sin x$ and graph y_1, y_2, and $y = y_1 + y_2$ on the same coordinate system.

5. We let $y_1 = 4$ and $y_2 = 2\sin x$ (which is a sine curve with an amplitude of 2 and a period of 2π). Then we graph y_1, y_2, and $y = y_1 + y_2$ on the same coordinate system.

9. We let $y_1 = \dfrac{1}{2}x$ and $y_2 = -\cos x$ (which is a cosine curve reflected about the x-axis). Then we graph y_1, y_2, and $y = y_1 + y_2$ on the same coordinate system.

13. We let $y_1 = 3\sin x$ (which is a sine curve with an amplitude of 3 and a period of 2π) and

$y_2 = \cos 2x$ (which is a cosine curve with an amplitude of 1 and a period or $\dfrac{2\pi}{2}$ or π).

Then we graph y_1, y_2, and $y = y_1 + y_2$ on the same coordinate system.

17. We let $y_1 = \sin x$ and $y_2 = \sin \dfrac{x}{2}$ (which is a sine curve with amplitude of 1 and a period of $\dfrac{2\pi}{1/2}$ or 4π). Then we graph y_1, y_2, and $y = y_1 + y_2$ on the same coordinate system.

21. We let $y_1 = \cos x$ and $y_2 = \dfrac{1}{2}\sin 2x$ (which is a sine curve with amplitude of $\dfrac{1}{2}$ and period of $\dfrac{2\pi}{2}$ or π). Then we graph y_1, y_2, and $y = y_1 + y_2$ on the same coordinate system.

25. $y = x \sin x$

x	0	$\frac{\pi}{2}$	π	$\frac{3\pi}{2}$	2π	$\frac{5\pi}{2}$	3π	$\frac{7\pi}{2}$	4π
$\sin x$	0	1	0	-1	0	1	0	-1	0
$x \sin x$	0	$\frac{\pi}{2}$	0	$-\frac{3\pi}{2}$	0	$\frac{5\pi}{2}$	0	$-\frac{7\pi}{2}$	0

29. $s = vt$

$= (20\ \dfrac{\text{ft}}{\text{sec}})\,(60\ \text{sec})$　　　We change 1 minute to 60 seconds so that units will agree

$= 1{,}200\ \text{ft}$

33. $120\ \text{rpm} = (120)(2\pi)$ radians per minute

$= 240\pi$ radians per minute

To convert this to radians per second, we divide by 60 because there are 60 seconds in 1 minute.

240π radians per minute $= \dfrac{240\pi}{60}$

$= 4\pi$ radians per second

Problem Set 4.6

1. We interchange x and y and then solve for y in terms of x:

$x = y^2 + 4$ or $y^2 + 4 = x$

$y^2 = x - 4$

$y = \pm\sqrt{x - 4}$

5.   We interchange x and y and then solve for y in terms of x:

$$x = 3y - 2 \text{ or } 3y - 2 = x$$
$$3y = x + 2$$
$$y = \frac{x + 2}{3}$$

9.   We interchange x and y and then solve for y in terms of x:

$$x = 3^y$$
$$y = \log_3 x$$

Then we graph these on the same coordinate axes.

$y = 3^x$			$x = 3^y$	
x	y		x	y
-2	$\frac{1}{9}$		$\frac{1}{9}$	-2
-1	$\frac{1}{3}$		$\frac{1}{3}$	-1
0	1		1	0
1	3		3	1
2	9		9	2

17.   By definition, the range for $y = \arccos x$ is $0 \leq y \leq \pi$.

21.   The angle between 0 and π whose cosine is -1 is π.

25.   The angle between 0 and π whose cosine is $-\dfrac{1}{\sqrt{2}}$ is $\dfrac{3\pi}{4}$.

29.   The angle between $-\dfrac{\pi}{2}$ and $\dfrac{\pi}{2}$ whose tangent is $\sqrt{3}$ is $\dfrac{\pi}{3}$.

33.   The angle between $-\dfrac{\pi}{2}$ and $\dfrac{\pi}{2}$ whose tangent is $-\dfrac{1}{\sqrt{3}}$ is $-\dfrac{\pi}{6}$.

37.   The angle between 0 and π whose cosine is $\dfrac{\sqrt{3}}{2}$ is $\dfrac{\pi}{6}$.

41.   Scientific Calculator: 0.8425 $\boxed{+/-}$ $\boxed{\text{inv}}$ $\boxed{\cos}$

Graphing Calculator: $\boxed{\text{2nd}}$ $\boxed{\cos}$ $\boxed{(}$ $\boxed{(-)}$ 0.8425 $\boxed{)}$ $\boxed{\text{ENTER}}$

Answer: $147.4°$

45. Scientific Calculator: 0.9627 | inv | | sin |

Graphing Calculator: | 2nd | | sin | | (| 0.9627 |) | | ENTER |

Answer: 74.3°

49. Scientific Calculator: 2.748 | +/− | | inv | | tan |

Graphing Calculator: | 2nd | | tan | | (| | (−) | 2.748 |) | | ENTER |

Answer: −70.0°

53. Let $\theta = \tan^{-1} \frac{3}{4}$, then $\tan \theta = \frac{3}{4}$ and $-\frac{\pi}{2} < \theta < \frac{\pi}{2}$

Next, we draw a triangle and find the hypotenuse:

$$\text{hypotenuse} = \sqrt{3^2 + 4^2}$$
$$= \sqrt{9 + 16}$$
$$= \sqrt{25}$$
$$= 5$$

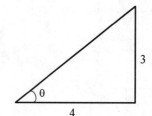

Then, we find $\cos \theta = \frac{4}{5}$.

57. Let $\theta = \cos^{-1} \frac{1}{\sqrt{5}}$, then $\cos \theta = \frac{1}{\sqrt{5}}$ and $0 \le \theta \le \pi$.

We are looking for $\sec \theta$.

$$\sec \theta = \frac{1}{\cos \theta}$$
$$= \frac{1}{1/\sqrt{5}}$$
$$= \sqrt{5}$$

61. Let $\theta = \tan^{-1} \frac{1}{2}$, then $\tan \theta = \frac{1}{2}$ and $-\frac{\pi}{2} < \theta < \frac{\pi}{2}$.

We want to find $\cot \theta = \frac{1}{\tan \theta} = \frac{1}{1/2} = 2$

65. Let $\theta = \cos^{-1} \frac{1}{2}$, then $\cos \theta = \frac{1}{2}$ and $0 \le \theta \le \pi$. We want to find $\cos \theta$ which is equal to $\frac{1}{2}$.

69. Let $\theta = \cos^{-1} x$, then $\cos \theta = x$ and $0 \le \theta \le \pi$. We want to find $\cos \theta$ which is equal to x.

73. Let $\theta = \tan^{-1} x$, then $\tan \theta = x$ and $-\frac{\pi}{2} < \theta < \frac{\pi}{2}$.

We draw a triangle and find the hypotenuse using the Pythagorean theorem:

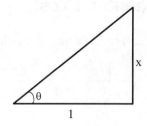

$$\text{hypotenuse} = \sqrt{x^2 + 1^2} = \sqrt{x^2 + 1}$$

From the figure, we find $\sin \theta = \dfrac{x}{\sqrt{x^2 + 1}}$

77. Let $\theta = \cos^{-1} \dfrac{1}{x}$, then $\cos \theta = \dfrac{1}{x}$ and $0 \le \theta \le \pi$.

We want to find $\sec \theta = \dfrac{1}{\cos \theta} = \dfrac{1}{1/x} = x$

81. The graph is a sine curve with amplitude of 2 and period of $\dfrac{2\pi}{\pi}$ or 2.

85. The graph is a standard sine curve with phase shift of $\dfrac{\pi}{4}$.

89. Amplitude $= 3$

Period $= \dfrac{2\pi}{2} = \pi$

Phase Shift $= \dfrac{\pi/3}{2} = \dfrac{\pi}{6}$

$A = \dfrac{\pi}{6}$

$E = \dfrac{\pi}{6} + \pi = \dfrac{7\pi}{6}$

$C = \dfrac{1}{2}(\dfrac{\pi}{6} + \dfrac{7\pi}{6}) = \dfrac{1}{2}(\dfrac{4\pi}{3}) = \dfrac{2\pi}{3}$

$B = \dfrac{1}{2}(\dfrac{\pi}{6} + \dfrac{2\pi}{3}) = \dfrac{1}{2}(\dfrac{5\pi}{6}) = \dfrac{5\pi}{12}$

$D = \dfrac{1}{2}(\dfrac{2\pi}{3} + \dfrac{7\pi}{6}) = \dfrac{1}{2}(\dfrac{11\pi}{6}) = \dfrac{11\pi}{12}$

The 5 points we use on the x-axis are $\dfrac{\pi}{6}, \dfrac{5\pi}{12}, \dfrac{2\pi}{3}, \dfrac{11\pi}{12}, \dfrac{7\pi}{6}$.

The 2 points we use on the y-axis are -3 and 3.

Chapter 4 Test

8. The x-values at which y is $\dfrac{1}{2}$ are

$$-\frac{11\pi}{3}, -\frac{7\pi}{3}, -\frac{5\pi}{3}, -\frac{\pi}{3}, \frac{\pi}{3}, \frac{5\pi}{3}, \frac{7\pi}{3}, \text{ and } \frac{11\pi}{3}.$$

9. Amplitude $= 1$

Period $= \dfrac{2\pi}{\pi} = 2$

10. Amplitude $= |-3| = 3$

Period $= 2\pi$

This graph is reflected across the x-axis.

11. Amplitude $= 3$

Period $= \dfrac{2\pi}{2} = \pi$

Vertical translation $= 2$

12. Amplitude $= 2$

Period $= \dfrac{2\pi}{\pi} = 2$

13. Amplitude $= 1$

Period $= 2\pi$

Phase shift $= -\dfrac{\pi}{4}$

$$A = -\frac{\pi}{4}$$

$$E = -\frac{\pi}{4} + 2\pi = \frac{7\pi}{4}$$

$$C = \frac{1}{2}\left(-\frac{\pi}{4} + \frac{7\pi}{4}\right) = \frac{1}{2}\left(\frac{3\pi}{2}\right) = \frac{3\pi}{4}$$

$$B = \frac{1}{2}\left(-\frac{\pi}{4} + \frac{3\pi}{4}\right) = \frac{1}{2}\left(\frac{\pi}{2}\right) = \frac{\pi}{4}$$

$$D = \frac{1}{2}\left(\frac{3\pi}{4} + \frac{7\pi}{4}\right) = \frac{1}{2}\left(\frac{5\pi}{2}\right) = \frac{5\pi}{4}$$

Points we use on the x-axis are $-\dfrac{\pi}{4}, \dfrac{\pi}{4}, \dfrac{3\pi}{4}, \dfrac{5\pi}{4}, \dfrac{7\pi}{4}$.

Points we use on the y-axis are -1 and 1.

14. Amplitude $= 1$

Period $= 2\pi$

Phase shift $= \dfrac{\pi}{2}$

$A = \dfrac{\pi}{2}$

$E = \dfrac{\pi}{2} + 2\pi = \dfrac{5\pi}{2}$

$C = \dfrac{1}{2}(\dfrac{\pi}{2} + \dfrac{5\pi}{2}) = \dfrac{1}{2}(3\pi) = \dfrac{3\pi}{2}$

$B = \dfrac{1}{2}(\dfrac{\pi}{2} + \dfrac{3\pi}{2}) = \dfrac{1}{2}(2\pi) = \pi$

$D = \dfrac{1}{2}(\dfrac{3\pi}{2} + \dfrac{5\pi}{2}) = \dfrac{1}{2}(4\pi) = 2\pi$

Points we use on the x-axis are $\dfrac{\pi}{2}, \pi, \dfrac{3\pi}{2}, 2\pi, \dfrac{5\pi}{2}$.

Points we use on the y-axis are -1 and 1.

15. Amplitude $= 3$

Period $= \dfrac{2\pi}{2} = \pi$

Phase shift $= \dfrac{\pi/3}{2} = \dfrac{\pi}{6}$

$A = \dfrac{\pi}{6}$

$E = \dfrac{\pi}{6} + \pi = \dfrac{7\pi}{6}$

$C = \dfrac{1}{2}(\dfrac{\pi}{6} + \dfrac{7\pi}{6}) = \dfrac{1}{2}(\dfrac{4\pi}{3}) = \dfrac{2\pi}{3}$

$B = \dfrac{1}{2}(\dfrac{\pi}{6} + \dfrac{2\pi}{3}) = \dfrac{1}{2}(\dfrac{5\pi}{6}) = \dfrac{5\pi}{12}$

$D = \dfrac{1}{2}(\dfrac{2\pi}{3} + \dfrac{7\pi}{6}) = \dfrac{1}{2}(\dfrac{11\pi}{6}) = \dfrac{11\pi}{12}$

Points we use on the x-axis are $\dfrac{\pi}{6}, \dfrac{5\pi}{12}, \dfrac{2\pi}{3}, \dfrac{11\pi}{12}, \dfrac{7\pi}{6}$.

Points we use on the y-axis are -3 and 3.

16. Amplitude $= 3$

Period $= \dfrac{2\pi}{\pi/3} = 6$

Phase shift $= \dfrac{\pi/3}{\pi/3} = 1$

Vertical translation $= -3$

$A = 1$

$E = 1 + 6 = 7$

$C = \dfrac{1}{2}(1+7) = \dfrac{1}{2}(8) = 4$

$B = \dfrac{1}{2}(1+4) = \dfrac{1}{2}(5) = \dfrac{5}{2}$

$D = \dfrac{1}{2}(4+7) = \dfrac{1}{2}(11) = \dfrac{11}{2}$

Points we use on the x-axis are $1, \dfrac{5}{2}, 4, \dfrac{11}{2}, 7$.

We move the graph of $y = 3\sin(\dfrac{\pi}{3}x - \dfrac{\pi}{3})$ down 3 units. Points we use on the y-axis are 0 and -6.

17. Using Problem 13 above, draw in the asymptotes where $x = 0$, at $-\dfrac{\pi}{4}, \dfrac{3\pi}{4}$, and $\dfrac{7\pi}{4}$.

Now sketch in the reciprocal function using the asymptotes and the curve

$y = \sin(x + \dfrac{\pi}{4})$ as your guides.

18. Period $= \dfrac{\pi}{2}$

Phase shift $= \dfrac{\pi/2}{2} = \dfrac{\pi}{4}$

Considering first the period of $\frac{\pi}{2}$, the asymptotes would be at $-\pi/4$ and $\pi/4$.

Since the graph has phase shift of $\pi/4$, the asymptotes are at $-\pi/4 + \dfrac{\pi}{4} = 0$

and $\dfrac{\pi}{4} + \dfrac{\pi}{4} = \dfrac{\pi}{2}$.

For one complete cycle, the asymptotes will be at 0 and $\pi/2$. The x-intercept will be at $\pi/4$.

19. Amplitude $= 2$

Period $= \dfrac{2\pi}{3}$

Phase shift $= \dfrac{\pi}{3}$

$$A = \frac{\pi}{3}$$

$$E = \frac{\pi}{3} + \frac{2\pi}{3} = \pi$$

$$C = \frac{1}{2}(\frac{\pi}{3} + \pi) = \frac{1}{2}(\frac{4\pi}{3}) = \frac{2\pi}{3}$$

$$B = \frac{1}{2}(\frac{\pi}{3} + \frac{2\pi}{3}) = \frac{1}{2}(\pi) = \frac{\pi}{2}$$

$$D = \frac{1}{2}(\frac{2\pi}{3} + \pi) = \frac{1}{2}(\frac{5\pi}{3}) = \frac{5\pi}{6}$$

Points we use on the x-axis are $\frac{\pi}{3}, \frac{\pi}{2}, \frac{2\pi}{3}, \frac{5\pi}{6}, \pi$.

Points we use on the y-axis are -2 and 2.

Then, extend the graph from $-\frac{\pi}{3}$ to $\frac{5\pi}{3}$.

20. Amplitude $= 2$

$$\text{Period} = \frac{2\pi}{\pi/2} = 4$$

$$\text{Phase shift} = \frac{\pi/4}{\pi/2} = \frac{1}{2}$$

$$A = \frac{1}{2}$$

$$E = \frac{1}{2} + 4 = \frac{9}{2}$$

$$C = \frac{1}{2}(\frac{1}{2} + \frac{9}{2}) = \frac{1}{2}(5) = \frac{5}{2}$$

$$B = \frac{1}{2}(\frac{1}{2} + \frac{5}{2}) = \frac{1}{2}(3) = \frac{3}{2}$$

$$D = \frac{1}{2}(\frac{5}{2} + \frac{9}{2}) = \frac{1}{2}(7) = \frac{7}{2}$$

Points we use on the x-axis are $\frac{1}{2}, \frac{3}{2}, \frac{5}{2}, \frac{7}{2}, \frac{9}{2}$.

Points we use on the y-axis are -2 and 2.

Then, extend the graph from $-\frac{1}{2}$ to $\frac{13}{2}$.

21. The graph is a sine curve with amplitude of 2, period of 4π, and phase shift of $-\pi$.

To find B, we use $\quad 4\pi = \frac{2\pi}{B}$

$$4\pi B = 2\pi$$

$$B = \frac{1}{2}$$

To find C, we use $\quad -\pi = -\dfrac{C}{B}$

$$-\pi = -\frac{C}{1/2}$$

$$\frac{\pi}{2} = C$$

Therefore, the equation is $y = 2\sin(\dfrac{1}{2}x + \dfrac{\pi}{2})$

22. The graph is a sine curve with amplitude of $\dfrac{1}{2}$ and period of 4, that has been translated up $\dfrac{1}{2}$ unit. There is no phase shift.

To find B, we use $\quad 4 = \dfrac{2\pi}{B}$

$$4B = 2\pi$$

$$B = \frac{\pi}{2}$$

Therefore, the equation is $y = \dfrac{1}{2} + \dfrac{1}{2}\sin\dfrac{\pi}{2}x$.

23. Let $y_1 = \frac{1}{2}x$ (a straight line) and $y_2 = -\sin x$ (a sine curve reflected about the x-axis). Then graph y_1, y_2, and $y = y_1 + y_2$ on the same coordinate system.

24. Let $y_1 = \sin x$ and $y_2 = \cos 2x$ (a cosine curve with period of π). Then graph y_1, y_2, and $y = y_1 + y_2$ on the same coordinate system.

25. The graph of $y = \cos^{-1} x$ is equivalent to $x = \cos y$ and $0 \le y \le \pi$. We make a table of values to aid in the graphing.

x	y
1	0
0	$\frac{\pi}{2}$
-1	π

26. The graph of $y = \arcsin x$ is equivalent to $x = \sin y$ and $-\pi/2 \le y \le \dfrac{\pi}{2}$. We make a table of values to aid in graphing:

x	y
-1	$-\dfrac{\pi}{2}$
0	0
1	$\dfrac{\pi}{2}$

27. The angle between $-\dfrac{\pi}{2}$ and $\dfrac{\pi}{2}$ whose sine is $\dfrac{1}{2}$ is $\dfrac{\pi}{6}$.

28. The angle between 0 and π whose cosine is $-\dfrac{\sqrt{3}}{2}$ is $\dfrac{5\pi}{6}$.

29. The angle between $-\dfrac{\pi}{2}$ and $\dfrac{\pi}{2}$ whose tangent is -1 is $-\dfrac{\pi}{4}$.

30. The angle between $-\dfrac{\pi}{2}$ and $\dfrac{\pi}{2}$ whose sine is 1 is $\dfrac{\pi}{2}$.

31. Scientific Calculator: 0.5934 | inv | | sin |

Graphing Calculator: | 2nd | | sin | | (| 0.5934 |) | | ENTER |

Answer: 36.4°

32. Scientific Calculator: 0.8302 | +/− | | inv | | tan |

Graphing Calculator: | 2nd | | tan | | (| | (−) | 0.8302 |) | | ENTER |

Answer: −39.7°

33. Scientific Calculator: 0.6981 | +/− | | inv | | cos |

Graphing Calculator: | 2nd | | cos | | (| | (−) | 0.6981 |) | | ENTER |

Answer: 134.3°

34. Scientific Calculator: 0.2164 | +/− | | inv | | sin |

Graphing Calculator: | 2nd | | sin | | (| | (−) | 0.2164 |) | | ENTER |

Answer: −12.5°

35. Let $\theta = \cos^{-1}\frac{2}{3}$, then $\cos\theta = \frac{2}{3}$ and $0 \le \theta \le \pi$.

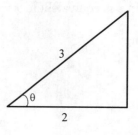

Next, we draw a triangle and find the opposite side:

$$\text{opposite side} = \sqrt{3^2 - 2^2}$$
$$= \sqrt{9 - 4}$$
$$= \sqrt{5}$$

From the figure, we find $\tan\theta = \frac{\sqrt{5}}{2}$.

36. Let $\theta = \tan^{-1}\frac{2}{3}$, then $\tan\theta = \frac{2}{3}$ and $-\frac{\pi}{2} < \theta < \frac{\pi}{2}$.

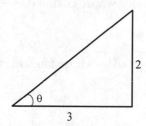

Next, we draw a triangle and find the hypotenuse:

$$\text{hypotenuse} = \sqrt{2^2 + 3^2}$$
$$= \sqrt{4 + 9}$$
$$= \sqrt{13}$$

From the figure, we find $\cos\theta = \dfrac{3}{\sqrt{13}}$

37. Let $\theta = \cos^{-1} x$, then $\cos\theta = x$ and $0 \le \theta \le \pi$.

Next, we draw a triangle and label the adjacent side and the hypotenuse.

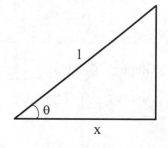

Now we find the opposite side using the Pythagorean theorem.

$$\text{Opposite side} = \sqrt{1^2 - x^2}$$
$$= \sqrt{1 - x^2}$$

From the figure, we find $\sin\theta = \dfrac{\sqrt{1-x^2}}{1}$
$$= \sqrt{1 - x^2}$$

38. Let $\theta = \sin^{-1} x$, then $\sin\theta = x$ and $-\frac{\pi}{2} \leq \theta \leq \frac{\pi}{2}$.

Next, we draw a triangle and label the opposite side and the hypotenuse.

Now we find the adjacent side using the Pythagorean theorem.

$$\text{adjacent side} = \sqrt{1^2 - x^2}$$
$$= \sqrt{1 - x^2}$$

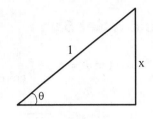

From the figure, we find $\tan\theta = \dfrac{x}{\sqrt{1 - x^2}}$

CHAPTER 5 Identities and Formulas

Problem Set 5.1

1.
$$\cos\theta\tan\theta = \cos\theta\cdot\frac{\sin\theta}{\cos\theta} \qquad \text{Ratio identity}$$

$$= \frac{\cos\theta\sin\theta}{\cos\theta} \qquad \text{Multiply}$$

$$= \sin\theta \qquad \text{Reduce}$$

5.
$$\frac{\tan A}{\sec A} = \frac{\frac{\sin A}{\cos A}}{\frac{1}{\cos A}} \qquad \text{Ratio identity and reciprocal identity}$$

$$= \frac{\sin A\cos A}{\cos A} \qquad \text{Divide}$$

$$= \sin A \qquad \text{Reduce}$$

9.
$$\cos x(\csc x + \tan x) = \cos x\csc x + \cos x\tan x \qquad \text{Distributive property}$$

$$= \cos x\cdot\frac{1}{\sin x} + \cos x\cdot\frac{\sin x}{\cos x} \qquad \text{Reciprocal and ratio identities}$$

$$= \frac{\cos x}{\sin x} + \frac{\cos x\sin x}{\cos x} \qquad \text{Multiply}$$

$$= \cot x + \sin x \qquad \text{Ratio identity and reduce second fraction}$$

13.
$$\cos^2 x(1 + \tan^2 x) = \cos^2 x(\sec^2 x) \qquad \text{Pythagorean identity}$$

$$= \cos^2 x\left(\frac{1}{\cos^2 x}\right) \qquad \text{Reciprocal identity}$$

$$= \frac{\cos^2 x}{\cos^2 x} \qquad \text{Multiply}$$

$$= 1 \qquad \text{Reduce}$$

17.
$$\frac{\cos^4 t - \sin^4 t}{\sin^2 t} = \frac{(\cos^2 t + \sin^2 t)(\cos^2 t - \sin^2 t)}{\sin^2 t} \qquad \text{Factor}$$

$$= \frac{1(\cos^2 t - \sin^2 t)}{\sin^2 t} \qquad \text{Pythagorean identity}$$

$$= \frac{\cos^2 t}{\sin^2 t} - \frac{\sin^2 t}{\sin^2 t} \qquad \text{Separate into 2 fractions}$$

$$= \cot^2 t - 1 \qquad \text{Ratio identity and reduce second fraction}$$

21. $\dfrac{1 - \sin^4\theta}{1 + \sin^2\theta} = \dfrac{(1 - \sin^2\theta)(1 + \sin^2\theta)}{1 + \sin^2\theta}$ Factor

$$= 1 - \sin^2\theta \qquad \text{Reduce}$$

$$= \cos^2\theta \qquad \text{Pythagorean identity}$$

25. $\sec^4\theta - \tan^4\theta = (\sec^2\theta - \tan^2\theta)(\sec^2\theta + \tan^2\theta)$ Factor

$$= 1(\sec^2\theta + \tan^2\theta) \qquad \text{Pythagorean identity}$$

$$= \frac{1}{\cos^2\theta} + \frac{\sin^2\theta}{\cos^2\theta} \qquad \text{Reciprocal and ratio identities}$$

$$= \frac{1 + \sin^2\theta}{\cos^2\theta} \qquad \text{Add fractions}$$

29. $\csc B - \sin B = \dfrac{1}{\sin B} - \sin B$ Reciprocal identity

$$= \frac{1}{\sin B} - \sin B \cdot \mathbf{\frac{\sin B}{\sin B}} \qquad \text{LCD is } \sin B$$

$$= \frac{1 - \sin^2 B}{\sin B} \qquad \text{Subtract fractions}$$

$$= \frac{\cos^2 B}{\sin B} \qquad \text{Pythagorean identity}$$

$$= \frac{\cos B}{\sin B} \cdot \cos B \qquad \text{Separate fractions}$$

$$= \cot B \cos B \qquad \text{Ratio identity}$$

33. $\dfrac{\cos x}{1 + \sin x} + \dfrac{1 + \sin x}{\cos x} = \dfrac{\cos x}{1 + \sin x} \cdot \mathbf{\dfrac{\cos x}{\cos x}} + \dfrac{1 + \sin x}{\cos x} \cdot \mathbf{\dfrac{1 + \sin x}{1 + \sin x}}$ LCD

$$= \frac{\cos^2 x}{\cos x(1 + \sin x)} + \frac{1 + 2\sin x + \sin^2 x}{\cos x(1 + \sin x)} \qquad \text{Multiply fractions}$$

$$= \frac{(\cos^2 x + \sin^2 x) + 1 + 2\sin x}{\cos x(1 + \sin x)} \qquad \text{Add fractions}$$

$$= \frac{1 + 1 + 2\sin x}{\cos x(1 + \sin x)} \qquad \text{Pythagorean identity}$$

$$= \frac{2 + 2\sin x}{\cos x(1 + \sin x)} \qquad \text{Add}$$

Continued on next page.

$$= \frac{2(1 + \sin x)}{\cos x(1 + \sin x)} \qquad \text{Factor out a 2}$$

$$= \frac{2}{\cos x} \qquad \text{Reduce}$$

$$= 2 \sec x \qquad \text{Reciprocal identity}$$

37. $\dfrac{1 - \sec x}{1 + \sec x} = \dfrac{1 - \frac{1}{\cos x}}{1 + \frac{1}{\cos x}}$ Reciprocal identity

$$= \frac{\cos x(1 - \frac{1}{\cos x})}{\cos x(1 + \frac{1}{\cos x})} \qquad \text{Multiply numerator and denominator by LCD}$$

$$= \frac{\cos x - 1}{\cos x + 1} \qquad \text{Distributive property}$$

41. $\dfrac{1 - \sin t}{1 + \sin t} = \dfrac{1 - \sin t}{1 + \sin t} \cdot \dfrac{1 - \sin t}{1 - \sin t}$ Multiply numerator and denominator by $1 - \sin t$

$$= \frac{(1 - \sin t)^2}{1 - \sin^2 t} \qquad \text{Multiply fractions}$$

$$= \frac{(1 - \sin t)^2}{\cos^2 t} \qquad \text{Pythagorean identity}$$

45. $(\sec x - \tan x)^2 = \left(\dfrac{1}{\cos x} - \dfrac{\sin x}{\cos x} \right)^2$ Reciprocal identity and ratio identity

$$= \left(\frac{1 - \sin x}{\cos x} \right)^2 \qquad \text{Subtract fractions}$$

$$= \frac{(1 - \sin x)^2}{\cos^2 x} \qquad \text{Property of exponents}$$

$$= \frac{(1 - \sin x)^2}{1 - \sin^2 x} \qquad \text{Pythagorean identity}$$

$$= \frac{(1 - \sin x)(1 - \sin x)}{(1 - \sin x)(1 + \sin x)} \qquad \text{Factor}$$

$$= \frac{1 - \sin x}{1 + \sin x} \qquad \text{Reduce}$$

49. $\dfrac{\sin x + 1}{\cos x + \cot x} = \dfrac{\sin x + 1}{\cos x + \frac{\cos x}{\sin x}}$ Ratio identity

$\qquad = \dfrac{\boldsymbol{\sin x}}{\boldsymbol{\sin x}} \cdot \dfrac{(\sin x + 1)}{\left(\cos x + \frac{\cos x}{\sin x}\right)}$ Multiply numerator and denominator by LCD

$\qquad = \dfrac{\sin x(\sin x + 1)}{\sin x \cos x + \cos x}$ Distributive property

$\qquad = \dfrac{\sin x(\sin x + 1)}{\cos x(\sin x + 1)}$ Factor

$\qquad = \dfrac{\sin x}{\cos x}$ Reduce

$\qquad = \tan x$ Ratio identity

53. $\dfrac{\sin^2 B - \tan^2 B}{1 - \sec^2 B} = \dfrac{\sin^2 B - \frac{\sin^2 B}{\cos^2 B}}{1 - \frac{1}{\cos^2 B}}$ Ratio identity and reciprocal identity

$\qquad = \dfrac{\boldsymbol{\cos^2 B}\left(\sin^2 B - \frac{\sin^2 B}{\cos^2 B}\right)}{\boldsymbol{\cos^2 B}\left(1 - \frac{1}{\cos^2 B}\right)}$ Multiply numerator and denominator by LCD

$\qquad = \dfrac{\cos^2 B \sin^2 B - \sin^2 B}{\cos^2 B - 1}$ Distributive property

$\qquad = \dfrac{\sin^2 B(\cos^2 B - 1)}{\cos^2 B - 1}$ Factor

$\qquad = \sin^2 B$ Reduce

57. $\dfrac{\sin^3 A - 8}{\sin A - 2} = \dfrac{(\sin A - 2)(\sin^2 A + 2\sin A + 4)}{\sin A - 2}$ Factor as difference of 2 cubes

$\qquad = \sin^2 A + 2\sin A + 4$ Reduce

61. $\dfrac{\sin^2 x + \sin x \cos x}{\cos x - 2\cos^3 x} = \dfrac{\sin x(\sin x + \cos x)}{\cos x(1 - 2\cos^2 x)}$ Factor

$\qquad = \dfrac{\sin x(\sin x + \cos x)}{\cos x[(1 - \cos^2 x) - \cos^2 x]}$ Regrouping denominator

$\qquad = \dfrac{\sin x(\sin x + \cos x)}{\cos x(\sin^2 x - \cos^2 x)}$ Pythagorean identity

Continued on next page.

$$= \frac{\sin x(\sin x + \cos x)}{\cos x(\sin x + \cos x)(\sin x - \cos x)} \qquad \text{Factor}$$

$$= \frac{\sin x/\cos x}{\sin x - \cos x} \qquad \text{Reduce}$$

$$= \frac{\tan x}{\sin x - \cos x} \qquad \text{Ratio identity}$$

65. $\quad \dfrac{1}{\tan\theta + \cot\theta} = \dfrac{1}{\frac{\sin\theta}{\cos\theta} + \frac{\cos\theta}{\sin\theta}} \qquad$ Ratio identities

$$= \frac{\sin\theta\cos\theta\,(1)}{\sin\theta\cos\theta\left(\frac{\sin\theta}{\cos\theta} + \frac{\cos\theta}{\sin\theta}\right)} \qquad \text{Multiply numerator and denominator by LCD}$$

$$= \frac{\sin\theta\cos\theta}{\sin^2\theta + \cos^2\theta} \qquad \text{Distributive property}$$

$$= \frac{\sin\theta\cos\theta}{1} \qquad \text{Pythagorean identity}$$

$$= \sin\theta\cos\theta \qquad \text{Divide}$$

69. $\quad \tan^2 x \cos x = (\sec^2 x - 1)(\cos x) \qquad$ Pythagorean identity

$$= (\sec x + 1)(\sec x - 1)\cos x \qquad \text{Factor}$$

$$= (\sec x + 1)[(\frac{1}{\cos x} - 1)\cos x] \qquad \text{Reciprocal identity}$$

$$= (1 + \sec x)(1 - \cos x) \qquad \text{Distributive property}$$

73. $\quad \dfrac{1 + \sin\phi}{1 - \sin\phi} - \dfrac{1 - \sin\phi}{1 + \sin\phi}$

$$= \frac{(1 + \sin\phi)(1 + \sin\phi)}{(1 - \sin\phi)(1 + \sin\phi)} - \frac{(1 - \sin\phi)(1 - \sin\phi)}{(1 + \sin\phi)(1 - \sin\phi)} \qquad \text{LCD}$$

$$= \frac{(1 + 2\sin\phi + \sin^2\phi) - (1 - 2\sin\phi + \sin^2\phi)}{(1 - \sin\phi)(1 + \sin\phi)} \qquad \text{Subtract fractions}$$

$$= \frac{4\sin\phi}{1 - \sin^2\phi} \qquad \text{Simplify}$$

$$= \frac{4\sin\phi}{\cos^2\phi} \qquad \text{Pythagorean identity}$$

$$= 4 \cdot \frac{\sin\phi}{\cos\phi} \cdot \frac{1}{\cos\phi} \qquad \text{Separate fraction}$$

$$= 4\tan\phi\sec\phi \qquad \text{Ratio and reciprocal identities}$$

77. If $\theta = 30°$, then $\sin 30° \neq \dfrac{1}{\cos 30°}$

$$\frac{1}{2} \neq \frac{1}{\sqrt{3}/2}$$

$$\frac{1}{2} \neq \frac{2}{\sqrt{3}}$$

Problem Set 5.2

1. $\sin 15° = \sin(45° - 30°)$

$$= \sin 45° \cos 30° - \cos 45° \sin 30°$$

$$= \left(\frac{\sqrt{2}}{2} \right) \left(\frac{\sqrt{3}}{2} \right) - \left(\frac{\sqrt{2}}{2} \right) \left(\frac{1}{2} \right)$$

$$= \frac{\sqrt{6}}{4} - \frac{\sqrt{2}}{4}$$

$$= \frac{\sqrt{6} - \sqrt{2}}{4}$$

5. $\sin \dfrac{7\pi}{12} = \sin \left(\dfrac{3\pi}{12} + \dfrac{4\pi}{12} \right)$

$$= \sin \left(\frac{\pi}{4} + \frac{\pi}{3} \right)$$

$$= \sin \frac{\pi}{4} \cos \frac{\pi}{3} + \cos \frac{\pi}{4} \sin \frac{\pi}{3}$$

$$= \left(\frac{\sqrt{2}}{2} \right) \left(\frac{1}{2} \right) + \left(\frac{\sqrt{2}}{2} \right) \left(\frac{\sqrt{3}}{2} \right)$$

$$= \frac{\sqrt{2}}{4} + \frac{\sqrt{6}}{4}$$

$$= \frac{\sqrt{2} + \sqrt{6}}{4}$$

9. $\sin(x + 2\pi) = \sin x \cos 2\pi + \cos x \sin 2\pi$

$$= (\sin x)(1) + (\cos x)(0)$$

$$= \sin x$$

13. $\cos(180° - \theta) = \cos 180° \cos \theta + \sin 180° \sin \theta$

$$= -1(\cos \theta) + 0(\sin \theta)$$

$$= -\cos \theta$$

17.
$$\tan(x + \frac{\pi}{4}) = \frac{\tan x + \tan \frac{\pi}{4}}{1 - \tan x \tan \frac{\pi}{4}}$$
$$= \frac{\tan x + 1}{1 - (\tan x)(1)}$$
$$= \frac{1 + \tan x}{1 - \tan x}$$

21. $\sin 3x \cos 2x + \cos 3x \sin 2x = \sin(3x + 2x) = \sin 5x$

25. $\cos 15° \cos 75° - \sin 15° \sin 75° = \cos(15° + 75°) = \cos 90° = 0$

29.
$$y = 3 \cos(7x) \cos(5x) + 3 \sin(7x) \sin(5x)$$
$$= 3[\cos(7x) \cos(5x) + \sin(7x) \sin(5x)]$$
$$= 3 \cos(7x - 5x)$$
$$y = 3 \cos 2x$$

The graph is a cosine curve with an amplitude of 3 and period of $\frac{2\pi}{2}$ or π.

33.
$$y = 2(\sin x \cos \frac{\pi}{3} + \cos x \sin \frac{\pi}{3})$$
$$y = 2 \sin(x + \frac{\pi}{3})$$

The graph is a sine curve with:

Amplitude $= 2$

Period $= 2\pi$

Phase shift $= -\frac{\pi}{3}$

$$A = -\frac{\pi}{3}$$
$$E = -\frac{\pi}{3} + 2\pi = \frac{5\pi}{3}$$
$$C = \frac{1}{2}(-\frac{\pi}{3} + \frac{5\pi}{3}) = \frac{1}{2}(\frac{4\pi}{3}) = \frac{2\pi}{3}$$
$$B = \frac{1}{2}(-\frac{\pi}{3} + \frac{2\pi}{3}) = \frac{1}{2}(\frac{\pi}{3}) = \frac{\pi}{6}$$
$$D = \frac{1}{2}(\frac{2\pi}{3} + \frac{5\pi}{3}) = \frac{1}{2}(\frac{7\pi}{3}) = \frac{7\pi}{6}$$

The 5 points on the x-axis we use are: $-\frac{\pi}{3}, \frac{\pi}{6}, \frac{2\pi}{3}, \frac{7\pi}{6}, \frac{5\pi}{3}$.

The 2 points on the y-axis we use are: 2 and -2.

37. If $\sin A = \dfrac{1}{\sqrt{5}}$ with A in QI, then

$$\cos A = \sqrt{1 - \sin^2 A}$$

$$= \sqrt{1 - \frac{1}{5}}$$

$$= \sqrt{\frac{4}{5}}$$

$$= \frac{2}{\sqrt{5}}$$

Also, $\tan A = \dfrac{\sin A}{\cos A}$

$$= \frac{\frac{1}{\sqrt{5}}}{\frac{2}{\sqrt{5}}}$$

$$= \frac{1}{2}$$

We have $\tan A = \dfrac{1}{2}$ and $\tan B = \dfrac{3}{4}$.

Therefore, $\tan(A + B) = \dfrac{\frac{1}{2} + \frac{3}{4}}{1 - (\frac{1}{2})(\frac{3}{4})}$

$$= \frac{\frac{5}{4}}{\frac{5}{8}}$$

$$= 2$$

$$\cot(A + B) = \frac{1}{\tan(A + B)}$$

$$= \frac{1}{2}$$

The angle $(A + B)$ terminates in QI because its tangent is positive. (If its tangent were negative, it would terminate in QII.)

41. $\sin 2x = \sin(x + x)$

$$= \sin x \cos x + \cos x \sin x$$

$$= 2 \sin x \cos x$$

45. $\cos(x - 90°) - \cos(x + 90°)$

$= [\cos x \cos 90° + \sin x \sin 90°] - [\cos x \cos 90° - \sin x \sin 90°]$ Sum and difference formulas

$= [(\cos x)(0) + (\sin x)(1)] - [(\cos x)(0) - (\sin x)(1)]$ Substitute exact values

$= \sin x - (-\sin x)$ Multiply

$= \sin x$ Subtract

49. $\cos(x + \dfrac{\pi}{4}) + \cos(x - \dfrac{\pi}{4})$

$= (\cos x \cos \dfrac{\pi}{4} - \sin x \sin \dfrac{\pi}{4}) + (\cos x \cos \dfrac{\pi}{4} + \sin x \sin \dfrac{\pi}{4})$ Sum and difference formulas

$= (\cos x) \dfrac{\sqrt{2}}{2} - (\sin x) \dfrac{\sqrt{2}}{2} + (\cos x) \dfrac{\sqrt{2}}{2} + (\sin x) \dfrac{\sqrt{2}}{2}$ Substitute exact values

$= \sqrt{2} \cos x$ Add

53. $\sin(A + B) + \sin(A - B)$

$= (\sin A \cos B + \cos A \sin B) + (\sin A \cos B - \cos A \sin B)$ Sum and difference formulas

$= 2 \sin A \cos B$ Add

57. $\sec(A + B) = \dfrac{1}{\cos(A + B)}$ Reciprocal identity

$= \dfrac{1}{\cos(A + B)} \cdot \dfrac{\mathbf{\cos(A - B)}}{\mathbf{\cos(A - B)}}$ Multiply numerator and denominator by $\cos(A - B)$

$= \dfrac{\cos(A - B)}{(\cos A \cos B - \sin A \sin B)(\cos A \cos B + \sin A \sin B)}$ Sum and difference identities

$= \dfrac{\cos(A - B)}{\cos^2 A \cos^2 B - \sin^2 A \sin^2 B}$ Multiply

$= \dfrac{\cos(A - B)}{\cos^2 A \cos^2 B + \sin^2 A \cos^2 B - \sin^2 A \sin^2 B - \sin^2 A \cos^2 B}$ Add and subtract $\sin^2 A \cos^2 B$ in denominator

$= \dfrac{\cos(A - B)}{\cos^2 B(\cos^2 A + \sin^2 A) - \sin^2 A(\sin^2 B + \cos^2 B)}$ Factor by grouping

$= \dfrac{\cos(A - B)}{\cos^2 B(1) - \sin^2 A(1)}$ Pythagorean identity

$= \dfrac{\cos(A - B)}{\cos^2 B - \sin^2 A}$ Multiply

61. The graph of $Y_1 = -\cos X$ is a cosine curve reflected across the x-axis.

The graph of $Y_2 = \cos(\pi - X)$ is equivalent to

$$Y_2 = \cos(-X + \pi)$$
$$= \cos[-(X - \pi)] \qquad \text{Factor}$$
$$= \cos(X - \pi) \qquad \text{Cosine is an even function}$$

The graphs of Y_1 and Y_2 are the same.

65. Amplitude $= 3$

Period $= \dfrac{2\pi}{1/2} = 4\pi$

69. Period $= \dfrac{2\pi}{3}$

Asymptotes occur where $y = \sin 3x$ is zero, or at 0, $\dfrac{\pi}{3}$, and $\dfrac{2\pi}{3}$.

We use the graph of $y = \sin 3x$ and the asymptotes to sketch the graph of $y = \csc 3x$.

73. Amplitude $= \dfrac{1}{2}$

Period $= \dfrac{2\pi}{\pi/2} = 4$

Problem Set 5.3

1. If $\sin A = -\dfrac{3}{5}$ with A in QIII, then

$$\cos A = -\sqrt{1 - \sin^2 A}$$
$$= -\sqrt{1 - \dfrac{9}{25}}$$
$$= -\sqrt{\dfrac{16}{25}}$$
$$= -\dfrac{4}{5}$$

Therefore, $\sin 2A = 2 \sin A \cos A$

$$= 2\left(-\dfrac{3}{5}\right)\left(-\dfrac{4}{5}\right)$$
$$= \dfrac{24}{25}$$

5. If $\cos x = \dfrac{1}{\sqrt{10}}$ with x in QIV, then

$$\cos 2x = 2\cos^2 x - 1$$

$$= 2\left(\dfrac{1}{\sqrt{10}}\right)^2 - 1$$

$$= 2\left(\dfrac{1}{10}\right) - 1$$

$$= \dfrac{1}{5} - 1$$

$$= -\dfrac{4}{5}$$

9. If $\tan\theta = \frac{5}{12}$ with θ in QI, we can draw the triangle at the right and find the hypotenuse using the Pythagorean theorem.

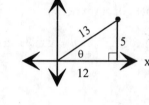

$$\begin{aligned} \text{hypotenuse} &= \sqrt{5^2 + 12^2} \\ &= \sqrt{25 + 144} \\ &= \sqrt{169} = 13 \end{aligned}$$

Then $\sin\theta = \frac{5}{13}$ and $\cos\theta = \frac{12}{13}$.

Therefore, $\sin 2\theta = 2\sin\theta\cos\theta$

$$= 2\left(\tfrac{5}{13}\right)\left(\tfrac{12}{13}\right) = \tfrac{120}{169}$$

13. If $\csc t = \sqrt{5}$ with t in QII, then $\sin t = \frac{1}{\sqrt{5}}$.

Therefore, $\cos 2t = 1 - 2\sin^2 t$

$$= 1 - 2\left(\dfrac{1}{\sqrt{5}}\right)^2$$

$$= 1 - \dfrac{2}{5}$$

$$= \dfrac{3}{5}$$

17. $y = 4 - 8\sin^2 x$

$y = 4(1 - 2\sin^2 x)$

$y = 4\cos 2x$

The graph is a cosine curve with amplitude $= 4$ and period $= \dfrac{2\pi}{2} = \pi$.

21. $y = 1 - 2\sin^2 2x$

$y = \cos 2(2x)$

$y = \cos 4x$

The graph is a cosine curve with amplitude $= 1$ and period $= \dfrac{2\pi}{4} = \dfrac{\pi}{2}$.

25. $\cos 120° = -\cos 60°$

$= -\dfrac{1}{2}$

$\cos^2 60° - \sin^2 60° = \left(\dfrac{1}{2}\right)^2 - \left(\dfrac{\sqrt{3}}{2}\right)^2$

$= \dfrac{1}{4} - \dfrac{3}{4}$

$= -\dfrac{1}{2}$

Therefore, they are equal.

29. $2\sin 15° \cos 15° = \sin 2(15°)$

$= \sin 30°$

$= \dfrac{1}{2}$

33. $\sin \dfrac{\pi}{12} \cos \dfrac{\pi}{12} = \dfrac{1}{2}(2 \sin \dfrac{\pi}{12} \cos \dfrac{\pi}{12})$

$= \dfrac{1}{2} \sin 2(\dfrac{\pi}{12})$

$= \dfrac{1}{2} \sin \dfrac{\pi}{6}$

$= \dfrac{1}{2} \left(\dfrac{1}{2}\right)$

$= \dfrac{1}{4}$

37. $(\sin x - \cos x)^2 = \sin^2 x - 2\sin x \cos x + \cos^2 x$ Expand

$= (\sin^2 x + \cos^2 x) - 2\sin x \cos x$ Commutative property

$= 1 - 2\sin x \cos x$ Pythagorean identity

$= 1 - \sin 2x$ Double-angle identity

41.
$$\frac{\sin 2\theta}{1 - \cos 2\theta} = \frac{2\sin\theta\cos\theta}{1 - (1 - 2\sin^2\theta)} \qquad \text{Double-angle identities}$$

$$= \frac{2\sin\theta\cos\theta}{2\sin^2\theta} \qquad \text{Subtract}$$

$$= \frac{\cos\theta}{\sin\theta} \qquad \text{Reduce}$$

$$= \cot\theta \qquad \text{Ratio identity}$$

45.
$$\sin 3\theta = \sin(2\theta + \theta) \qquad \text{Addition}$$

$$= \sin 2\theta\cos\theta + \cos 2\theta\sin\theta \qquad \text{Sum formula}$$

$$= (2\sin\theta\cos\theta)\cos\theta + (1 - 2\sin^2\theta)\sin\theta \qquad \text{Double-angle identities}$$

$$= 2\sin\theta\cos^2\theta + \sin\theta - 2\sin^3\theta \qquad \text{Multiply}$$

$$= 2\sin\theta(1 - \sin^2\theta) + \sin\theta - 2\sin^3\theta \qquad \text{Pythagorean identity}$$

$$= 2\sin\theta - 2\sin^3\theta + \sin\theta - 2\sin^3\theta \qquad \text{Multiply}$$

$$= 3\sin\theta - 4\sin^3\theta \qquad \text{Add like terms}$$

49.
$$\frac{\cos 2\theta}{\sin\theta\cos\theta} = \frac{\cos^2\theta - \sin^2\theta}{\sin\theta\cos\theta} \qquad \text{Double-angle identity}$$

$$= \frac{\cos^2\theta}{\sin\theta\cos\theta} - \frac{\sin^2\theta}{\sin\theta\cos\theta} \qquad \text{Separate into 2 fractions}$$

$$= \frac{\cos\theta}{\sin\theta} - \frac{\sin\theta}{\cos\theta} \qquad \text{Reduce}$$

$$= \cot\theta - \tan\theta \qquad \text{Ratio identities}$$

53.
$$\frac{\cos B - \sin B\tan B}{\sec B\sin 2B} = \frac{\cos B - \sin B\left(\frac{\sin B}{\cos B}\right)}{\frac{1}{\cos B}\sin 2B} \qquad \text{Ratio identity and reciprocal identity}$$

$$= \frac{\mathbf{\cos B}\left(\cos B - \frac{\sin^2 B}{\cos B}\right)}{\mathbf{\cos B}\left(\frac{\sin 2B}{\cos B}\right)} \qquad \text{Multiply numerator and denominator by LCD}$$

$$= \frac{\cos^2 B - \sin^2 B}{\sin 2B} \qquad \text{Multiply}$$

$$= \frac{\cos 2B}{\sin 2B} \qquad \text{Double-angle identity}$$

$$= \cot 2B \qquad \text{Ratio identity}$$

57. $\dfrac{1 - \tan x}{1 + \tan x} = \dfrac{1 - \frac{\sin x}{\cos x}}{1 + \frac{\sin x}{\cos x}}$ Ratio identity

$\qquad = \dfrac{\cos x(1 - \frac{\sin x}{\cos x})}{\cos x(1 + \frac{\sin x}{\cos x})}$ Multiply numerator and denominator by LCD

$\qquad = \dfrac{\cos x - \sin x}{\cos x + \sin x}$ Distributive property

$\qquad = \dfrac{\cos x - \sin x}{\cos x + \sin x} \cdot \dfrac{\cos x - \sin x}{\cos x - \sin x}$ Multiply by a fraction equal to 1

$\qquad = \dfrac{\cos^2 x - 2\sin x \cos x + \sin^2 x}{\cos^2 x - \sin^2 x}$ Multiply fractions

$\qquad = \dfrac{1 - 2\sin x \cos x}{\cos^2 x - \sin^2 x}$ Pythagorean identity

$\qquad = \dfrac{1 - \sin 2x}{\cos 2x}$ Double-angle identities

61. First, we find θ: $\qquad x = 3\sin\theta$

$$\sin\theta = \frac{x}{3}$$

$$\theta = \arcsin\frac{x}{3}$$

Next, we find $\sin 2\theta$ by drawing the triangle at the right
and using the Pythagorean theorem:

$$\text{adjacent side} = \sqrt{3^2 - x^2}$$

$$= \sqrt{9 - x^2}$$

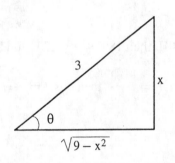

Then, $\sin 2\theta = 2\sin\theta \cos\theta$

$$= 2\left(\frac{x}{3}\right)\left(\frac{\sqrt{9 - x^2}}{3}\right)$$

$$= \frac{2}{9}x\sqrt{9 - x^2}$$

Last, we evaluate the given expression:

$$\frac{\theta}{2} - \frac{\sin 2\theta}{4} = \frac{1}{2}\arcsin\frac{x}{3} - \frac{1}{4}\left(\frac{2}{9}x\sqrt{9 - x^2}\right)$$

$$= \frac{1}{2}\arcsin\frac{x}{3} - \frac{1}{18}x\sqrt{9 - x^2}$$

$$= \frac{1}{2}\left(\arcsin\frac{x}{3} - \frac{x\sqrt{9 - x^2}}{9}\right)$$

65. Let $y_1 = 3$ (a horizontal line) and $y_2 = -3\cos x$ (a cosine curve reflected about the x-axis with amplitude of 3). Graph y_1, y_2, and $y = y_1 + y_2$ on the same coordinate system.

69. Let $y_1 = \dfrac{1}{2}x$ (a line) and $y_2 = \sin \pi x$ (a sine curve with amplitude $= 1$ and period $= \dfrac{2\pi}{\pi} = 2$). Graph y_1, y_2, and $y = y_1 + y_2$ on the same coordinate system.

Problem Set 5.4

1. If A is in QIV, then $270° < A < 360°$ and $135° < \dfrac{A}{2} < 180°$. Therefore, $\dfrac{A}{2}$ is in QII.

$$\sin \frac{A}{2} = \sqrt{\frac{1 - \cos A}{2}}$$

$$= \sqrt{\frac{1 - 1/2}{2}}$$

$$= \sqrt{\frac{1}{4}}$$

$$= \frac{1}{2}$$

5. If A is in QIII, then $180° < A < 270°$ and $90° < \dfrac{A}{2} < 135°$. Therefore, $\dfrac{A}{2}$ is in QII.

$$\text{If } \sin A = -\frac{3}{5}, \text{ then } \cos A = -\sqrt{1 - \sin^2 A}$$

$$= -\sqrt{1 - \frac{9}{25}}$$

$$= -\sqrt{\frac{16}{25}}$$

$$= -\frac{4}{5}$$

$$\text{Therefore, } \cos \frac{A}{2} = -\sqrt{\frac{1 + \cos A}{2}}$$

$$= -\sqrt{\frac{1 - 4/5}{2}}$$

$$= -\sqrt{\frac{1/5}{2}}$$

$$= -\sqrt{\frac{1}{10}}$$

$$= -\frac{1}{\sqrt{10}}$$

9. If B is in QIII, then $\dfrac{B}{2}$ is in QII. (See Problem 5 above.)

If $\sin B = -\dfrac{1}{3}$, then $\cos B = -\sqrt{1 - \sin^2 B}$

$$= -\sqrt{1 - \frac{1}{9}}$$

$$= -\sqrt{\frac{8}{9}}$$

$$= -\frac{2\sqrt{2}}{3}$$

Therefore, $\sin \dfrac{B}{2} = \sqrt{\dfrac{1 - \cos B}{2}}$

$$= \sqrt{\frac{1 + \frac{2\sqrt{2}}{3}}{2}} \qquad \text{(Multiply numerator and denominator by 3)}$$

$$= \sqrt{\frac{3 + 2\sqrt{2}}{6}}$$

13. Using the information from Problem 9 above:

$$\tan \frac{B}{2} = \frac{1 - \cos A}{\sin A}$$

$$= \frac{1 + \frac{2\sqrt{2}}{3}}{-\frac{1}{3}}$$

$$= \frac{3 + 2\sqrt{2}}{-1}$$

$$= -3 - 2\sqrt{2}$$

17. $\cos 2A = 1 - 2 \sin^2 A$

$$= 1 - 2(\frac{4}{5})^2$$

$$= 1 - \frac{32}{25}$$

$$= -\frac{7}{25}$$

21. If B is in QI, then $\dfrac{B}{2}$ must be in QI.

If $\sin B = \dfrac{3}{5}$, then $\cos B = \sqrt{1 - \sin^2 B}$

$$= \sqrt{1 - \frac{9}{25}} = \sqrt{\frac{16}{25}} = \frac{4}{5}$$

Therefore, $\cos\dfrac{B}{2} = \sqrt{\dfrac{1 + \cos B}{2}}$

$$= \sqrt{\frac{1 + \frac{4}{5}}{2}} = \sqrt{\frac{9}{10}} = \frac{3}{\sqrt{10}}$$

25. If $\sin A = \dfrac{4}{5}$, then $\cos A = -\sqrt{1 - \sin^2 A}$

$$= \sqrt{1 - \frac{16}{25}} = -\sqrt{\frac{9}{25}} = -\frac{3}{5}$$

We have $\quad \sin A = \dfrac{4}{5} \qquad\qquad \sin B = \dfrac{3}{5}$

$$\cos A = -\frac{3}{5} \qquad \cos B = \frac{4}{5} \text{ (from Problem 21)}$$

Therefore, $\cos(A - B) = \cos A \cos B + \sin A \sin B$

$$= -\frac{3}{5}\left(\frac{4}{5}\right) + \frac{4}{5}\left(\frac{3}{5}\right)$$

$$= -\frac{12}{25} + \frac{12}{25}$$

$$= 0$$

29. $y = 2\cos^2\dfrac{x}{2}$

$$y = 2\left(\pm\sqrt{\frac{1 + \cos x}{2}}\right)^2$$

$$y = 2\left(\frac{1 + \cos x}{2}\right)$$

$$y = 1 + \cos x$$

This graph is the standard cosine curve with a vertical translation of 1.

33. $\sin 75° = \sin\left(\dfrac{150°}{2}\right)$

$$= \sqrt{\frac{1 - \cos 150°}{2}}$$

$$= \sqrt{\frac{1 - (-\frac{\sqrt{3}}{2})}{2}}$$

$$= \sqrt{\frac{2 + \sqrt{3}}{4}}$$

$$= \frac{\sqrt{2 + \sqrt{3}}}{2}$$

37. $\dfrac{\csc\theta - \cot\theta}{2\csc\theta} = \dfrac{\dfrac{1}{\sin\theta} - \dfrac{\cos\theta}{\sin\theta}}{\dfrac{2}{\sin\theta}}$ Reciprocal and ratio identities

$$= \dfrac{\dfrac{1 - \cos\theta}{\sin\theta}}{\dfrac{2}{\sin\theta}}$$ Subtract

$$= \frac{1 - \cos\theta}{2}$$ Divide

$$= \sin^2\frac{\theta}{2}$$ Half-angle formula

41. $\tan\dfrac{B}{2} = \dfrac{1 - \cos B}{\sin B}$ Half-angle formula

$$= \frac{1}{\sin B} - \frac{\cos B}{\sin B}$$ Separate fractions

$$= \csc B - \cot B$$ Reciprocal and ratio identities

45. $\dfrac{\tan\theta + \sin\theta}{2\tan\theta} = \dfrac{\dfrac{\sin\theta}{\cos\theta} + \sin\theta}{\dfrac{2\sin\theta}{\cos\theta}}$ Ratio identity

$$= \frac{\sin\theta + \sin\theta\cos\theta}{2\sin\theta}$$ Multiply numerator and denominator by $\cos\theta$

$$= \frac{\sin\theta(1 + \cos\theta)}{2\sin\theta}$$ Factor

$$= \frac{1 + \cos\theta}{2}$$ Reduce

$$= \cos^2\frac{\theta}{2}$$ Half-angle formula

49. Let $\theta = \arcsin \frac{3}{5}$. Then $\sin \theta = \frac{3}{5}$ and $-\pi/2 \leq \theta \leq \pi/2$.

We want to find $\sin \theta$ which is equal to $\dfrac{3}{5}$.

53. Let $\theta = \tan^{-1} x$. Then $\tan \theta = \dfrac{x}{1}$ and $-\pi/2 < \theta < \pi/2$.

Next we draw a triangle and label the opposite and adjacent sides. Then we find the hypotenuse using the Pythagorean theorem:
$$\text{hypotenuse} = \sqrt{x^2 + 1^2} = \sqrt{x^2 + 1}$$
Using the figure, $\sin \theta = \dfrac{x}{\sqrt{x^2 + 1}}$

53. Let $\theta = \tan^{-1} x$. Then $\tan \theta = \dfrac{x}{1}$ and $-\pi/2 < \theta < \pi/2$.

Next we draw a triangle and label the opposite and adjacent sides. Then we find the hypotenuse using the Pythagorean theorem:
$$\text{hypotenuse} = \sqrt{x^2 + 1^2} = \sqrt{x^2 + 1}$$
Using the figure, $\sin \theta = \dfrac{x}{\sqrt{x^2 + 1}}$

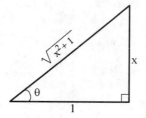

57. The graph of $y = \sin^{-1} x$ is equivalent to $x = \sin y$. Choose values for y and find the corresponding x-values.

Problem Set 5.5

1. Let $\alpha = \arcsin \frac{3}{5}$ and $\beta = \arctan 2$.

Then $\sin \alpha = \frac{3}{5}$ and $-90° \leq \alpha \leq 90°$ and $\tan \beta = 2$ and $-90° < \beta < 90°$

Also, $\sin \left(\arcsin \frac{3}{5} - \arctan 2 \right) = \sin (\alpha - \beta)$
$$= \sin \alpha \cos \beta - \cos \alpha \sin \beta$$

Drawing and labeling a triangle for α and another for β we have

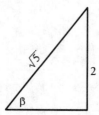 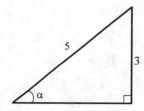

Using the Pythagorean theorem we find the missing sides:

$$\text{adjacent side} = \sqrt{5^2 - 3^2}$$

$$= \sqrt{25 - 9}$$

$$= \sqrt{16}$$

$$= 4$$

Then, $\sin \alpha = \frac{3}{5}$

$$\cos \alpha = \frac{4}{5}$$

$$\text{adjacent side} = \sqrt{(\sqrt{5})^2 - 2^2}$$

$$= \sqrt{5 - 4}$$

$$= 1$$

$$\sin \beta = \frac{2}{\sqrt{5}}$$

$$\cos \beta = \frac{1}{\sqrt{5}}$$

Substituting these above, we get

$$\sin \alpha \cos \beta - \cos \alpha \sin \beta = \frac{3}{5}\left(\frac{1}{\sqrt{5}}\right) - \frac{4}{5}\left(\frac{2}{\sqrt{5}}\right)$$

$$= \frac{3}{5\sqrt{5}} - \frac{8}{5\sqrt{5}}$$

$$= -\frac{5}{5\sqrt{5}}$$

$$= -\frac{1}{\sqrt{5}}$$

5. Let $\alpha = \cos^{-1}\frac{1}{\sqrt{5}}$. Then $\cos \alpha = \frac{1}{\sqrt{5}}$ and $0° \leq \alpha \leq 180°$.

Also, $\sin\left(2\cos^{-1}\frac{1}{\sqrt{5}}\right) = \sin 2\alpha$

$$= 2\sin \alpha \cos \alpha$$

We draw and label the sides of a triangle and using the

Pythagorean theorem, we find the adjacent side:

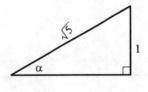

$$\text{adjacent side} = \sqrt{(\sqrt{5})^2 - 1^2} = \sqrt{4} = 2$$

From the figure, we have

$$\sin \alpha = \frac{1}{\sqrt{5}} \quad \text{and} \quad \cos \alpha = \frac{2}{\sqrt{5}}$$

Substituting these above, we get

$$2\sin \alpha \cos \alpha = 2\left(\frac{1}{\sqrt{5}}\right)\left(\frac{2}{\sqrt{5}}\right) = \frac{4}{5}$$

9. Let $\alpha = \sin^{-1} x$. Then $\sin \alpha = \dfrac{x}{1}$ and $-90° \le \alpha \le 90°$.

Also, $\sin(2 \sin^{-1} x) = \sin 2\alpha$

$$= 2 \sin \alpha \cos \alpha$$

We draw and label the sides of a triangle and using
the Pythagorean theorem, we find the adjacent side:

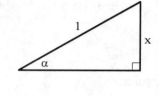

$$\text{adjacent side} = \sqrt{1^2 - x^2} = \sqrt{1 - x^2}$$

From the figure, we have

$$\sin \alpha = x \quad \text{and} \quad \cos \alpha = \sqrt{1 - x^2}$$

Substituting these above, we get

$$2 \sin \alpha \cos \alpha = 2x\sqrt{1 - x^2}$$

13. $\sin 30° \sin 120° = \frac{1}{2}[\cos(-90°) - \cos 150°]$

$$\frac{1}{2} \cdot \frac{\sqrt{3}}{2} = \frac{1}{2}\left[0 - \left(-\frac{\sqrt{3}}{2} \right) \right]$$

$$\frac{\sqrt{3}}{4} = \frac{\sqrt{3}}{4}$$

17. $\cos 8x \cos 2x = \dfrac{1}{2}[\cos(8x + 2x) + \cos(8x - 2x)]$

$$= \frac{1}{2}(\cos 10x + \cos 6x)$$

21. $\sin 4\pi \sin 2\pi = \dfrac{1}{2}[\cos(4\pi - 2\pi) - \cos(4\pi + 2\pi)]$

$$= \frac{1}{2}[\cos 2\pi - \cos 6\pi]$$

$$= \frac{1}{2}[1 - 1]$$

$$= 0$$

25. $\sin 7x + \sin 3x = 2 \sin \dfrac{7x + 3x}{2} \cos \dfrac{7x - 3x}{2}$

$$= 2 \sin 5x \cos 2x$$

29. $\sin\dfrac{7\pi}{12} - \sin\dfrac{\pi}{12} = 2\cos\dfrac{7\pi/12+\pi/12}{2}\sin\dfrac{7\pi/12-\pi/12}{2}$

$$= 2\cos\dfrac{\pi}{3}\sin\dfrac{\pi}{4}$$

$$= 2\left(\dfrac{1}{2}\right)\left(\dfrac{1}{\sqrt{2}}\right)$$

$$= \dfrac{1}{\sqrt{2}}$$

33. $\dfrac{\sin 4x + \sin 6x}{\cos 4x - \cos 6x} = \dfrac{2\sin 5x\cos(-x)}{-2\sin 5x\sin(-x)}$ Sum to product formulas

$$= -\dfrac{\cos(-x)}{\sin(-x)} \qquad\qquad\text{Reduce}$$

$$= -\dfrac{\cos x}{-\sin x} \qquad\qquad\text{Cosine is an even function; sine is an odd function}$$

$$= \cot x \qquad\qquad\text{Ratio identity}$$

37. The graph is a sine curve with a phase shift of $-\dfrac{\pi}{4}$.

41. Amplitude $= 1$

Period $= \dfrac{2\pi}{2} = \pi$

Phase shift $= \dfrac{\pi/2}{2} = \dfrac{\pi}{4}$

The 5 points we use on the x-axis are:

$$A = \dfrac{\pi}{4}$$

$$E = \dfrac{\pi}{4} + \pi = \dfrac{5\pi}{4}$$

$$C = \dfrac{1}{2}\left(\dfrac{\pi}{4} + \dfrac{5\pi}{4}\right) = \dfrac{3\pi}{4}$$

$$B = \dfrac{1}{2}\left(\dfrac{\pi}{4} + \dfrac{3\pi}{4}\right) = \dfrac{\pi}{2}$$

$$D = \dfrac{1}{2}\left(\dfrac{3\pi}{4} + \dfrac{5\pi}{4}\right) = \pi$$

The 2 points we use on the y-axis are: 1 and -1.

45. Amplitude $= 3$

Period $= \dfrac{2\pi}{\pi} = 2$

Phase shift $= \dfrac{\pi/2}{\pi} = \dfrac{1}{2}$

The 5 points we use on the x-axis are:

$A = \dfrac{1}{2}$

$E = \dfrac{1}{2} + 2 = \dfrac{5}{2}$

$C = \dfrac{1}{2}\left(\dfrac{1}{2} + \dfrac{5}{2}\right) = \dfrac{3}{2}$

$B = \dfrac{1}{2}\left(\dfrac{1}{2} + \dfrac{3}{2}\right) = 1$

$D = \dfrac{1}{2}\left(\dfrac{3}{2} + \dfrac{5}{2}\right) = 2$

The 2 points we use on the y-axis are: 3 and -3.

Chapter 5 Test

1. $\sin\theta \sec\theta = \sin\theta \cdot \dfrac{1}{\cos\theta}$ Reciprocal identity

$\qquad\qquad = \dfrac{\sin\theta}{\cos\theta}$ Multiply

$\qquad\qquad = \tan\theta$ Ratio identity

2. $\dfrac{\cot\theta}{\csc\theta} = \dfrac{\frac{\cos\theta}{\sin\theta}}{\frac{1}{\sin\theta}}$ Ratio and reciprocal identities

$\qquad\quad = \cos\theta$ Divide

3. $(\sec x - 1)(\sec x + 1) = \sec^2 x - 1$ Multiply

$\qquad\qquad\qquad\qquad\quad = \tan^2 x$ Pythagorean identity

4. $\tan\theta \sin\theta = \dfrac{\sin\theta}{\cos\theta} \sin\theta$ Ratio identity

$\qquad\qquad = \dfrac{\sin^2\theta}{\cos\theta}$ Multiply

$\qquad\qquad = \dfrac{1 - \cos^2\theta}{\cos\theta}$ Pythagorean identity

$\qquad\qquad = \dfrac{1}{\cos\theta} - \dfrac{\cos^2\theta}{\cos\theta}$ Separate into 2 fractions

$\qquad\qquad = \sec\theta - \cos\theta$ Reciprocal identity and reduce second fraction

5. $\dfrac{\cos t}{1 - \sin t} = \dfrac{\cos t}{1 - \sin t} \cdot \dfrac{1 + \sin t}{1 + \sin t}$ Multiply by a fraction equal to 1

$\qquad\qquad = \dfrac{\cos t(1 + \sin t)}{1 - \sin^2 t}$ Multiply

$\qquad\qquad = \dfrac{\cos t\,(1 + \sin t)}{\cos^2 t}$ Pythagorean identity

$\qquad\qquad = \dfrac{1 + \sin t}{\cos t}$ Reduce

6.

$$\frac{1}{1-\sin t} + \frac{1}{1+\sin t} = \frac{1(1+\sin t) + 1(1-\sin t)}{(1-\sin t)(1+\sin t)} \qquad \text{Add fractions}$$

$$= \frac{2}{1-\sin^2 t} \qquad \text{Simplify}$$

$$= \frac{2}{\cos^2 t} \qquad \text{Pythagorean identity}$$

$$= 2\sec^2 t \qquad \text{Reciprocal identity}$$

7.

$$\sin(\theta - 90°) = \sin\theta \cos 90° - \cos\theta \sin 90° \qquad \text{Difference formula}$$

$$= (\sin\theta)(0) - (\cos\theta)(1) \qquad \text{Substitute exact values}$$

$$= -\cos\theta \qquad \text{Simplify}$$

8.

$$\cos(\frac{\pi}{2} + \theta) = \cos\frac{\pi}{2}\cos\theta - \sin\frac{\pi}{2}\sin\theta \qquad \text{Sum formula}$$

$$= 0(\cos\theta) - 1(\sin\theta) \qquad \text{Substitute exact values}$$

$$= -\sin\theta \qquad \text{Simplify}$$

9.

$$\cos^4 A - \sin^4 A = (\cos^2 A + \sin^2 A)(\cos^2 A - \sin^2 A) \qquad \text{Factor}$$

$$= 1(\cos^2 A - \sin^2 A) \qquad \text{Pythagorean identity}$$

$$= \cos 2A \qquad \text{Double-angle formula}$$

10.

$$\frac{\sin 2A}{1 - \cos 2A} = \frac{2\sin A \cos A}{1 - (1 - 2\sin^2 A)} \qquad \text{Double-angle formulas}$$

$$= \frac{2\sin A \cos A}{2\sin^2 A} \qquad \text{Subtract}$$

$$= \frac{\cos A}{\sin A} \qquad \text{Reduce}$$

$$= \cot A \qquad \text{Ratio identity}$$

11.

$$\frac{\cos 2x}{\sin x \cos x} = \frac{\cos^2 x - \sin^2 x}{\sin x \cos x} \qquad \text{Double-angle formula}$$

$$= \frac{\cos^2 x}{\sin x \cos x} - \frac{\sin^2 x}{\sin x \cos x} \qquad \text{Separate into 2 fractions}$$

$$= \frac{\cos x}{\sin x} - \frac{\sin x}{\cos x} \qquad \text{Reduce}$$

$$= \cot x - \tan x \qquad \text{Ratio identities}$$

12. $\dfrac{\tan x}{\sec x + 1} = \dfrac{\frac{\sin x}{\cos x}}{\frac{1}{\cos x} + 1}$ Ratio and reciprocal identities

$\qquad\qquad = \dfrac{\sin x}{1 + \cos x}$ Multiply numerator and denominator by $\cos x$

$\qquad\qquad = \tan \dfrac{x}{2}$ Half-angle formula

13. A is in QIV. Then, $\cos A = \sqrt{1 - \sin^2 A}$

$$= \sqrt{1 - \frac{9}{25}}$$

$$= \sqrt{\frac{16}{25}}$$

$$= \frac{4}{5}$$

B is in QII. Then, $\cos B = -\sqrt{1 - \sin^2 A}$

$$= -\sqrt{1 - \frac{144}{169}}$$

$$= -\sqrt{\frac{25}{169}}$$

$$= -\frac{5}{13}$$

We have $\qquad \sin A = -\dfrac{3}{5} \qquad\qquad \sin B = \dfrac{12}{13}$

$$\cos A = \frac{4}{5} \qquad\qquad \cos B = -\frac{5}{13}$$

Therefore, $\sin(A + B) = \sin A \cos B + \cos A \sin B$

$$= -\frac{3}{5}\left(-\frac{5}{13}\right) + \frac{4}{5}\left(\frac{12}{13}\right)$$

$$= \frac{15}{65} + \frac{48}{65}$$

$$= \frac{63}{65}$$

14. $\cos(A - B) = \cos A \cos B + \sin A \sin B$

$$= \frac{4}{5}\left(-\frac{5}{13}\right) + \left(-\frac{3}{5}\right)\left(\frac{12}{13}\right) \qquad \text{(See Problem 13)}$$

$$= -\frac{20}{65} - \frac{36}{65}$$

$$= -\frac{56}{65}$$

15. $\cos 2B = 1 - 2\sin^2 B$

$$= 1 - 2(\frac{12}{13})^2$$

$$= 1 - 2(\frac{144}{169})$$

$$= 1 - \frac{288}{169}$$

$$= -\frac{119}{169}$$

16. $\sin 2B = 2\sin B \cos B$

$$= 2(\frac{12}{13})(-\frac{5}{13}) \qquad \text{(See Problem 13)}$$

$$= -\frac{120}{169}$$

17. Using the information from Problem 13 above:

If $270° \leq A \leq 360°$, then $135° \leq \frac{A}{2} \leq 180°$. Therefore, $\frac{A}{2}$ is in QII.

$$\sin \frac{A}{2} = \sqrt{\frac{1 - \cos A}{2}}$$

$$= \sqrt{\frac{1 - \frac{4}{5}}{2}}$$

$$= \sqrt{\frac{1}{10}}$$

$$= \frac{1}{\sqrt{10}}$$

18. $\cos \frac{A}{2} = -\sqrt{\frac{1 + \cos A}{2}} \qquad \text{(See Problem 17)}$

$$= -\sqrt{\frac{1 + 4/5}{2}}$$

$$= -\sqrt{\frac{9}{10}}$$

$$= -\frac{3}{\sqrt{10}}$$

19. $\sin 75° = \sin(30° + 45°)$

$$= \sin 30° \cos 45° + \cos 30° \sin 45°$$

$$= \frac{1}{2}\left(\frac{\sqrt{2}}{2}\right) + \frac{\sqrt{3}}{2}\left(\frac{\sqrt{2}}{2}\right)$$

$$= \frac{\sqrt{2}}{4} + \frac{\sqrt{6}}{4}$$

$$= \frac{\sqrt{2} + \sqrt{6}}{4}$$

20. $\cos 15° = \cos(45° - 30°)$

$$= \cos 45° \cos 30° + \sin 45° \sin 30°$$

$$= \frac{\sqrt{2}}{2}\left(\frac{\sqrt{3}}{2}\right) + \frac{\sqrt{2}}{2}\left(\frac{1}{2}\right)$$

$$= \frac{\sqrt{6}}{4} + \frac{\sqrt{2}}{4}$$

$$= \frac{\sqrt{6} + \sqrt{2}}{4}$$

21. $\tan \dfrac{\pi}{12} = \tan\left(\dfrac{\pi}{3} - \dfrac{\pi}{4}\right)$

$$= \frac{\tan \frac{\pi}{3} - \tan \frac{\pi}{4}}{1 + \tan \frac{\pi}{3} \tan \frac{\pi}{4}}$$

$$= \frac{\sqrt{3} - 1}{1 + (\sqrt{3})(1)}$$

$$= \frac{\sqrt{3} - 1}{\sqrt{3} + 1}$$

22. $\cot \dfrac{\pi}{12} = \dfrac{1}{\tan \pi/12}$

$$= \frac{1}{\left[\frac{\sqrt{3}-1}{\sqrt{3}+1}\right]} \qquad \text{(See Problem 21)}$$

$$= \frac{\sqrt{3} + 1}{\sqrt{3} - 1}$$

23. $\cos 4x \cos 5x - \sin 4x \sin 5x = \cos(4x + 5x)$

$$= \cos 9x$$

24.
$$\sin 15° \cos 75° + \cos 15° \sin 75° = \sin(15° + 75°)$$
$$= \sin 90°$$
$$= 1$$

25.
$$\cos 2A = 1 - 2\sin^2 A$$
$$= 1 - 2\left(-\frac{1}{\sqrt{5}}\right)^2$$
$$= 1 - \frac{2}{5}$$
$$= \frac{3}{5}$$

For $\cos\dfrac{A}{2}$, we must find $\cos A$.

$$\cos A = -\sqrt{1 - \sin^2 A}$$
$$= -\sqrt{1 - \frac{1}{5}} = -\sqrt{\frac{4}{5}} = -\frac{2}{\sqrt{5}} = -\frac{2\sqrt{5}}{5}$$

If $180° \le A \le 270°$, then $90° \le \dfrac{A}{2} \le 135°$, and $\dfrac{A}{2}$ is in QII.

$$\cos\frac{A}{2} = -\sqrt{\frac{1 + \cos A}{2}}$$
$$= -\sqrt{\frac{1 + \left(-\frac{2\sqrt{5}}{5}\right)}{2}}$$
$$= \sqrt{\frac{5 - 2\sqrt{5}}{10}}$$

26. If $\sec A = \sqrt{10}$, then $\cos A = \dfrac{1}{\sqrt{10}}$.

$$\sin A = \sqrt{1 - \cos^2 A}$$
$$= \sqrt{1 - \frac{1}{10}} = \sqrt{\frac{9}{10}} = \frac{3}{\sqrt{10}}$$

Then, $\sin 2A = 2\sin A \cos A$
$$= 2\left(\frac{3}{\sqrt{10}}\right)\left(\frac{1}{\sqrt{10}}\right) = \frac{6}{10} = \frac{3}{5}$$

Also, $\sin \dfrac{A}{2} = \sqrt{\dfrac{1 - \cos A}{2}}$

$$= \sqrt{\dfrac{1 - \frac{1}{\sqrt{10}}}{2}}$$

$$= \sqrt{\dfrac{\sqrt{10} - 1}{2\sqrt{10}}} \quad \text{or} \quad \sqrt{\dfrac{10 - \sqrt{10}}{20}}$$

27. $\tan(A + B) = \dfrac{\tan A + \tan B}{1 - \tan A \tan B}$

$$3 = \dfrac{\tan A + \frac{1}{2}}{1 - (\tan A)(\frac{1}{2})}$$

$$3 = \dfrac{2\tan A + 1}{2 - \tan A}$$

$$3(2 - \tan A) = 2\tan A + 1$$

$$6 - 3\tan A = 2\tan A + 1$$

$$-5\tan A = -5$$

$$\tan A = 1$$

28. $\cos 2x = 2\cos^2 x - 1$ Double-angle formula

$\quad \dfrac{1}{2} = 2\cos^2 x - 1$ Substitute given value

$\quad \dfrac{3}{2} = 2\cos^2 x$ Subtract 1 from both sides

$\cos^2 x = \dfrac{3}{4}$ Multiply both sides by $\dfrac{1}{2}$

$\cos x = \pm\dfrac{\sqrt{3}}{2}$ Take square root of both sides

29. Let $\alpha = \arcsin\frac{4}{5}$ and $\beta = \arctan 2$.

Then $\sin\alpha = \frac{4}{5}$ and $\tan\beta = \frac{2}{1}$.

We can draw 2 triangles and label the sides accordingly:

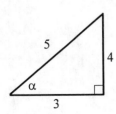 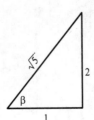

Continued on next page

From the figures on the previous page, we have

$$\sin\alpha = \frac{4}{5} \quad \sin\beta = \frac{2}{\sqrt{5}}$$

$$\cos\alpha = \frac{3}{5} \quad \cos\beta = \frac{1}{\sqrt{5}}$$

Therefore, $\cos(\alpha - \beta) = \cos\alpha\cos\beta + \sin\alpha\sin\beta$

$$= \frac{3}{5}\left(\frac{1}{\sqrt{5}}\right) + \frac{4}{5}\left(\frac{2}{\sqrt{5}}\right)$$

$$= \frac{3}{5\sqrt{5}} + \frac{8}{5\sqrt{5}}$$

$$= \frac{11}{5\sqrt{5}}$$

30. Let $\alpha = \arccos\dfrac{4}{5}$ and $\beta = \arctan 2$. Then $\cos\alpha = \dfrac{4}{5}$ and $\tan\beta = \dfrac{2}{1}$.

We can draw 2 triangles and label the sides accordingly:

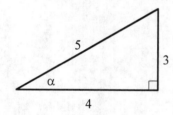
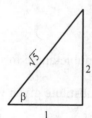

From the figures we have

$$\sin\alpha = \frac{3}{5} \quad \sin\beta = \frac{2}{\sqrt{5}}$$

$$\cos\alpha = \frac{4}{5} \quad \cos\beta = \frac{1}{\sqrt{5}}$$

Therefore, $\sin(\alpha + \beta) = \sin\alpha\cos\beta + \cos\alpha\sin\beta$

$$= \frac{3}{5}\left(\frac{1}{\sqrt{5}}\right) + \frac{4}{5}\left(\frac{2}{\sqrt{5}}\right)$$

$$= \frac{3}{5\sqrt{5}} + \frac{8}{5\sqrt{5}}$$

$$= \frac{11}{5\sqrt{5}}$$

31. Let $\alpha = \sin^{-1}x$. Then $\sin\alpha = x$

$$\cos 2\alpha = 1 - 2\sin^2\alpha$$
$$= 1 - 2x^2$$

32. Let $\alpha = \cos^{-1}x$. Then $\cos\alpha = \frac{x}{1}$. We can draw a triangle and use the Pythagorean theorem to find the opposite side:

$$\text{opposite side} = \sqrt{1^2 - x^2} = \sqrt{1 - x^2}$$

Therefore, $\sin\alpha = \dfrac{\sqrt{1-x^2}}{1} = \sqrt{1-x^2}$

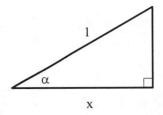

$$\sin 2\alpha = 2\sin\alpha\cos\alpha$$
$$= 2\sqrt{1-x^2}(x) = 2x\sqrt{1-x^2}$$

33. $\sin 6x\sin 4x = \dfrac{1}{2}[\cos(6x - 4x) - \cos(6x + 4x)]$

$$= \frac{1}{2}(\cos 2x - \cos 10x)$$

34. $\cos 15° + \cos 75° = 2\cos\dfrac{15° + 75°}{2}\cos\dfrac{15° - 75°}{2}$

$$= 2\cos 45° \cos(-30°)$$
$$= 2\cos 45° \cos 30°$$
$$= 2\left(\frac{\sqrt{2}}{2}\right)\left(\frac{\sqrt{3}}{2}\right)$$
$$= \frac{\sqrt{6}}{2}$$

CHAPTER 6 Equations

Problem Set 6.1

1. $2 \sin \theta = 1$

$\sin \theta = \dfrac{1}{2}$ Divide both sides by 2

$\theta = 30° \text{ or } 150°$ $\widehat{\theta} = 30°$ and θ is in QI or QII

5. $2 \tan \theta + 2 = 0$

$2 \tan \theta = -2$ Subtract 2 from both sides

$\tan \theta = -1$ Divide both sides by 2

$\theta = 135° \text{ or } 315°$ $\widehat{\theta} = 45°$ and θ is in QII or QIV

9. $2 \cos t = 6 \cos t - \sqrt{12}$

$-4 \cos t = -2\sqrt{3}$ Subtract $6 \cos t$ from both sides

$\cos t = \dfrac{\sqrt{3}}{2}$ Divide both sides by -4

$t = \dfrac{\pi}{6} \text{ or } \dfrac{11\pi}{6}$ $\widehat{t} = \dfrac{\pi}{6}$ and t is in QI or QIV

13. $4 \sin \theta - 3 = 0$

$4 \sin \theta = 3$ Add 3 to both sides

$\sin \theta = 0.75$ Divide both sides by 4

$\theta = 48.6° \text{ or } 131.4°$ $\widehat{\theta} = 48.6°$ and θ is in QI or QII

17. $\sin \theta - 3 = 5 \sin \theta$

$-3 = 4 \sin \theta$ Subtract $\sin \theta$ from both sides

$\sin \theta = -0.75$ Divide both sides by 4

$\theta = 228.6° \text{ or } 311.4°$ $\widehat{\theta} = 48.6°$ and θ is in QIII or QIV

21. $\tan x(\tan x - 1) = 0$

$\tan x = 0 \text{ or } \tan x - 1 = 0$ Set each factor $= 0$

$x = 0, \pi$ $\tan x = 1$ Solve each resulting equation

$$x = \dfrac{\pi}{4}, \dfrac{5\pi}{4}$$

25.

$$2\sin^2 x - \sin x - 1 = 0 \qquad \text{Standard form}$$

$$(2\sin x + 1)(\sin x - 1) = 0 \qquad \text{Factor}$$

$$2\sin x + 1 = 0 \text{ or } \sin x - 1 = 0 \qquad \text{Set each factor} = 0$$

$$2\sin x = -1 \qquad \sin x = 1 \qquad \text{Solve each resulting equation}$$

$$\sin x = -\frac{1}{2} \qquad x = \frac{\pi}{2}$$

$$x = \frac{7\pi}{6} \text{ or } \frac{11\pi}{6}$$

29.

$$\sqrt{3}\tan\theta - 2\sin\theta\tan\theta = 0 \qquad \text{Standard form}$$

$$\tan\theta(\sqrt{3} - 2\sin\theta) = 0 \qquad \text{Factor}$$

$$\tan\theta = 0 \text{ or } \quad \sqrt{3} - 2\sin\theta = 0 \qquad \text{Set each factor} = 0$$

$$\theta = 0° \text{ or } 180° \quad -2\sin\theta = -\sqrt{3} \qquad \text{Solve each resulting equation}$$

$$\sin\theta = \frac{\sqrt{3}}{2}$$

$$\theta = 60° \text{ or } 120°$$

33. $\quad 2\sin^2\theta - 2\sin\theta - 1 = 0 \qquad$ where $a = 2, b = -2, c = -1$

$$\sin\theta = \frac{-(-2) \pm \sqrt{(-2)^2 - 4(2)(-1)}}{2(2)}$$

$$= \frac{2 \pm \sqrt{12}}{4}$$

$$= \frac{2 \pm 2\sqrt{3}}{4}$$

$$= \frac{1 \pm \sqrt{3}}{2}$$

$$= \frac{1 \pm 1.7321}{2}$$

$\sin\theta = 1.3666 \quad$ or $\quad \sin\theta = -0.3660$

No solution $\qquad \widehat{\theta} = \mathbf{21.5°}$ and θ is in QIII or QIV

$$\theta = 201.5° \text{ or } 338.5°$$

37. $\quad 2\sin^2\theta + 1 = 4\sin\theta$

$$2\sin^2\theta - 4\sin\theta + 1 = 0$$

$$a = 2, \quad b = -4, \quad c = 1$$

Continued on next page

$$\sin\theta = \frac{-(-4) \pm \sqrt{(-4)^2 - 4(2)(1)}}{2(2)}$$

$$= \frac{4 \pm \sqrt{8}}{4}$$

$$= \frac{4 \pm 2\sqrt{2}}{4}$$

$$= \frac{2 \pm \sqrt{2}}{2}$$

$$= \frac{2 \pm 1.4142}{2}$$

$\sin\theta = 1.7071$ or $\sin\theta = 0.2929$

No solution $\theta = 17.0°$ or $163.0°$ $\widehat{\theta} = 17.0°$ and θ is in QI or QII

41. $4\sin t - \sqrt{3} = 2\sin t$

       $-\sqrt{3} = -2\sin t$ Subtract $4\sin t$ from both sides

         $\sin t = \dfrac{\sqrt{3}}{2}$ Divide both sides by -2

           $t = \dfrac{\pi}{3} + 2k\pi$ $\widehat{t} = \dfrac{\pi}{3}$ and t is in QI or QII

      or $t = \dfrac{2\pi}{3} + 2k\pi$

45.   In Problem 13 we found that $\theta = 48.6°$ or $131.4°$.

     Therefore, $\theta = 48.6° + 360°k$ or

           $\theta = 131.4° + 360°k$

49. $\sin(3A + 30°) = \dfrac{1}{2}$

    $3A + 30° = 30° + 360°k$ or $3A + 30° = 150° + 360°k$

       $3A = 360°k$ $3A = 120° + 360°k$

         $A = 120°k$ $A = 40° + 120°k$

53. $\sin(5A + 15°) = -\dfrac{1}{\sqrt{2}}$

    $5A + 15° = 225° + 360°k$ or $5A + 15° = 315° + 360°k$

       $5A = 210° + 360°k$ $5A = 300° + 360°k$

         $A = 42° + 72°k$ $A = 60° + 72°k$

57. $h = -16t^2 + vt \sin\theta$ where $t = 2$, $v = 1{,}500$, and $\theta = 30°$

$$= -16(2)^2 + (1{,}500)(2) \sin 30°$$

$$= -64 + 3{,}000\left(\frac{1}{2}\right)$$

$$= 1{,}436 \text{ ft}$$

65. $\sin(\theta + 45°) = \sin\theta \cos 45° + \cos\theta \sin 45°$

$$= \sin\theta\left(\frac{1}{\sqrt{2}}\right) + \cos\theta\left(\frac{1}{\sqrt{2}}\right)$$

$$= \frac{1}{\sqrt{2}} \sin\theta + \frac{1}{\sqrt{2}} \cos\theta$$

69. $\dfrac{1 - \tan^2 x}{1 + \tan^2 x} = \dfrac{1 - \frac{\sin^2 x}{\cos^2 x}}{1 + \frac{\sin^2 x}{\cos^2 x}}$ Ratio identity

$$= \frac{\cos^2 x - \sin^2 x}{\cos^2 x + \sin^2 x}$$ Multiply numerator and denominator by $\cos^2 x$

$$= \frac{\cos 2x}{1}$$ Double-angle formula and Pythagorean identity

$$= \cos 2x$$ Simplify

Problem Set 6.2

1. $\sqrt{3} \sec\theta = 2$

$\sec\theta = \dfrac{2}{\sqrt{3}}$ Divide both sides by $\sqrt{3}$

$\cos\theta = \dfrac{\sqrt{3}}{2}$ $\cos\theta = \dfrac{1}{\sec\theta}$

$\theta = 30°$ or $330°$ $\hat{\theta} = 30°$ and θ is in QI or QIV

5. $4 \sin\theta - 2 \csc\theta = 0$

$4 \sin\theta - \dfrac{2}{\sin\theta} = 0$ $\csc\theta = \dfrac{1}{\sin\theta}$

$4 \sin^2\theta - 2 = 0$ Multiply both sides by $\sin\theta$

$4 \sin^2\theta = 2$ Add 2 to both sides

Continued on next page.

$$\sin^2\theta = \frac{1}{2}$$ Divide both sides by 4

$$\sin\theta = \pm\frac{1}{\sqrt{2}}$$ Take square root of both sides

$$\theta = 45°, 135°, 225°, 315°$$ $\widehat{\theta} = 45°$

9.
$$\sin 2\theta - \cos\theta = 0$$

$$2\sin\theta\cos\theta - \cos\theta = 0$$ $\sin 2\theta = 2\sin\theta\cos\theta$

$$\cos\theta(2\sin\theta - 1) = 0$$ Factor out $\cos\theta$

$$\cos\theta = 0 \quad \text{or} \quad 2\sin\theta - 1 = 0$$ Set each factor $= 0$

$$\theta = 90°, 270° \quad\quad 2\sin\theta = 1$$ Solve each equation

$$\sin\theta = \frac{1}{2}$$

$$\theta = 30° \text{ or } 150°$$

13.
$$\cos 2x - 3\sin x - 2 = 0$$

$$1 - 2\sin^2 x - 3\sin x - 2 = 0$$ $\cos 2x = 1 - 2\sin^2 x$

$$2\sin^2 x + 3\sin x + 1 = 0$$ Multiply both sides by -1 and simplify

$$(2\sin x + 1)(\sin x + 1) = 0$$ Factor

$$2\sin x + 1 = 0 \quad \text{or} \quad \sin x + 1 = 0$$ Set each factor $= 0$

$$2\sin x = -1 \quad\quad\quad \sin x = -1$$ Solve each equation

$$\sin x = -\frac{1}{2} \quad\quad\quad x = \frac{3\pi}{2}$$

$$x = \frac{7\pi}{6} \text{ or } \frac{11\pi}{6}$$

17.
$$2\cos^2 x + \sin x - 1 = 0$$

$$2(1 - \sin^2 x) + \sin x - 1 = 0$$ $\cos^2 x = 1 - \sin^2 x$

$$2 - 2\sin^2 x + \sin x - 1 = 0$$ Simplify

$$2\sin^2 x - \sin x - 1 = 0$$ Multiply both sides by -1 and simplify

$$(2\sin x + 1)(\sin x - 1) = 0$$ Factor

$$2\sin x + 1 = 0 \quad \text{or} \quad \sin x - 1 = 0$$ Set each factor $= 0$

$$2\sin x = -1 \quad\quad\quad \sin x = 1$$ Solve each equation

$$\sin x = -\frac{1}{2} \quad\quad\quad x = \frac{\pi}{2}$$

$$x = \frac{7\pi}{6} \text{ or } \frac{11\pi}{6}$$

21. $2\sin x + \cot x - \csc x = 0$

$$2\sin x + \frac{\cos x}{\sin x} - \frac{1}{\sin x} = 0 \qquad\qquad \cot x = \frac{\cos x}{\sin x} \text{ and } \csc x = \frac{1}{\sin x}$$

$$2\sin^2 x + \cos x - 1 = 0 \qquad\qquad \text{Multiply both sides by } \sin x$$

$$2(1 - \cos^2 x) + \cos x - 1 = 0 \qquad\qquad \sin^2 x = 1 - \cos^2 x$$

$$2 - 2\cos^2 x + \cos x - 1 = 0 \qquad\qquad \text{Simplify}$$

$$2\cos^2 x - \cos x - 1 = 0 \qquad\qquad \text{Multiply both sides by } -1 \text{ and simplify}$$

$$(2\cos x + 1)(\cos x - 1) = 0 \qquad\qquad \text{Factor}$$

$$2\cos x + 1 = 0 \quad \text{or} \quad \cos x - 1 = 0 \qquad\qquad \text{Set each factor} = 0$$

$$2\cos x = -1 \qquad\qquad \cos x = 1 \qquad\qquad \text{Solve each equation and check}$$

$$\cos x = -\tfrac{1}{2} \qquad\qquad x = 0 \text{ which is not possible because cotangent and}$$

$$\text{cosecant are not defined when } x = 0$$

$$x = \frac{2\pi}{3} \text{ or } \frac{4\pi}{3}$$

25. $\sqrt{3}\sin\theta + \cos\theta = \sqrt{3}$

$$\cos\theta = \sqrt{3} - \sqrt{3}\sin\theta \qquad\qquad \text{Subtract } \sqrt{3}\sin\theta \text{ from both sides}$$

$$\cos\theta = \sqrt{3}(1 - \sin\theta) \qquad\qquad \text{Factor out } \sqrt{3}$$

$$\cos^2\theta = 3(1 - 2\sin\theta + \sin^2\theta) \qquad\qquad \text{Square both sides}$$

$$1 - \sin^2\theta = 3 - 6\sin\theta + 3\sin^2\theta \qquad\qquad \cos^2\theta = 1 - \sin^2\theta$$

$$4\sin^2\theta - 6\sin\theta + 2 = 0 \qquad\qquad \text{Simplify}$$

$$2\sin^2\theta - 3\sin\theta + 1 = 0 \qquad\qquad \text{Divide both sides by 2}$$

$$(2\sin\theta - 1)(\sin\theta - 1) = 0 \qquad\qquad \text{Factor}$$

$$2\sin\theta - 1 = 0 \quad \text{or} \quad \sin\theta - 1 = 0 \qquad\qquad \text{Set each factor} = 0$$

$$2\sin\theta = 1 \qquad\qquad \sin\theta = 1 \qquad\qquad \text{Solve each equation}$$

$$\sin\theta = \frac{1}{2}$$

$$\theta = 30° \text{ or } 150° \qquad \theta = 90° \qquad\qquad \text{Possible solutions}$$

Check each possible solution:

$$\sqrt{3}\sin 30° + \cos 30° \overset{?}{=} \sqrt{3} \qquad\qquad \sqrt{3}\sin 150° + \cos 150° \overset{?}{=} \sqrt{3}$$

$$\sqrt{3}\left(\frac{1}{2}\right) + \frac{\sqrt{3}}{2} \overset{?}{=} \sqrt{3} \qquad\qquad \sqrt{3}\left(\frac{1}{2}\right) + \left(-\frac{\sqrt{3}}{2}\right) \overset{?}{=} \sqrt{3}$$

$$\sqrt{3} = \sqrt{3} \qquad\qquad\qquad 0 \neq \sqrt{3}$$

Continued on next page.

$$\sqrt{3}\sin 90° + \cos 90° \overset{?}{=} \sqrt{3}$$

$$\sqrt{3}(1) + 0 \overset{?}{=} \sqrt{3}$$

$$\sqrt{3} = \sqrt{3}$$

Answers: 30° or 90°

29. $\qquad \sin\dfrac{\theta}{2} - \cos\theta = 0$

$$\sin\dfrac{\theta}{2} = \cos\theta \qquad\qquad\qquad \text{Add } \cos\theta \text{ to both sides}$$

$$\sin^2\dfrac{\theta}{2} = \cos^2\theta \qquad\qquad\qquad \text{Square both sides}$$

$$\dfrac{1-\cos\theta}{2} = \cos^2\theta \qquad\qquad\qquad \sin\dfrac{\theta}{2} = \pm\sqrt{\dfrac{1-\cos\theta}{2}}$$

$$1-\cos\theta = 2\cos^2\theta \qquad\qquad\qquad \text{Multiply both sides by 2}$$

$$2\cos^2\theta + \cos\theta - 1 = 0 \qquad\qquad \text{Rewrite in standard form}$$

$$(2\cos\theta - 1)(\cos\theta + 1) = 0 \qquad\qquad \text{Factor}$$

$$2\cos\theta - 1 = 0 \quad \text{or} \quad \cos\theta + 1 = 0 \qquad \text{Set each factor} = 0$$

$$2\cos\theta = 1 \qquad\qquad \cos\theta = -1 \qquad \text{Solve each equation and check}$$

$$\cos\theta = \tfrac{1}{2}$$

$$\theta = 60° \text{ or } 300° \qquad \theta = 180° \qquad \text{Possible solutions}$$

Check: $\qquad \sin\dfrac{60°}{2} - \cos 60° \overset{?}{=} 0 \qquad\qquad \sin\dfrac{300°}{2} - \cos 300° \overset{?}{=} 0$

$$\sin 30° - \cos 60° \overset{?}{=} 0 \qquad\qquad \sin 150° - \cos 300° \overset{?}{=} 0$$

$$\dfrac{1}{2} - \dfrac{1}{2} \overset{?}{=} 0 \qquad\qquad\qquad \dfrac{1}{2} - \dfrac{1}{2} \overset{?}{=} 0$$

$$0 = 0 \qquad\qquad\qquad\qquad 0 = 0$$

$$\sin\dfrac{180°}{2} - \cos 180° \overset{?}{=} 0$$

$$\sin 90° - \cos 180° \overset{?}{=} 0$$

$$1 - (-1) \overset{?}{=} 0$$

$$2 \neq 0$$

Answers: 60° or 300°

33.
$$6\cos\theta + 7\tan\theta = \sec\theta$$

$$6\cos\theta + \frac{7\sin\theta}{\cos\theta} = \frac{1}{\cos\theta} \qquad \tan\theta = \frac{\sin\theta}{\cos\theta} \text{ and } \sec\theta = \frac{1}{\cos\theta}$$

$$6\cos^2\theta + 7\sin\theta = 1 \qquad \text{Multiply both sides by } \cos\theta$$

$$6(1 - \sin^2\theta) + 7\sin\theta = 1 \qquad \cos^2\theta = 1 - \sin^2\theta$$

$$6 - 6\sin^2\theta + 7\sin\theta = 1 \qquad \text{Simplify}$$

$$-6\sin^2\theta + 7\sin\theta + 5 = 0 \qquad \text{Subtract 1 from both sides}$$

$$6\sin^2\theta - 7\sin\theta - 5 = 0 \qquad \text{Multiply both sides by } -1$$

$$(3\sin\theta - 5)(2\sin\theta + 1) = 0 \qquad \text{Factor}$$

$$3\sin\theta - 5 = 0 \quad \text{or} \quad 2\sin\theta + 1 = 0 \qquad \text{Set each factor} = 0$$

$$3\sin\theta = 5 \qquad 2\sin\theta = -1 \qquad \text{Solve each equation}$$

$$\sin\theta = \frac{5}{3} \qquad \sin\theta = -\frac{1}{2}$$

$$\text{No solution} \qquad \theta = 210° \text{ or } 330°$$

37.
$$7\sin^2\theta - 9\cos 2\theta = 0$$

$$7\sin^2\theta - 9(1 - 2\sin^2\theta) = 0 \qquad \cos 2\theta = 1 - 2\sin^2\theta$$

$$7\sin^2\theta - 9 + 18\sin^2\theta = 0 \qquad \text{Simplify left side}$$

$$25\sin^2\theta = 9 \qquad \text{Add 9 to both sides}$$

$$\sin^2\theta = \frac{9}{25} \qquad \text{Divide both sides by 25}$$

$$\sin\theta = \pm\frac{3}{5} \qquad \text{Take square root of both sides}$$

$$\sin\theta = \pm 0.6 \qquad \text{Divide right side}$$

$$\theta = 36.9°, 143.1°, \qquad \widehat{\theta} = 36.9° \text{ and } \theta \text{ is in QI,}$$
$$216.9°, \text{ or } 323.1° \qquad \text{QII, QIII, or QIV}$$

41. In Problem 23 we get $x = \dfrac{\pi}{4}$. All solutions would be $x = \dfrac{\pi}{4} + 2k\pi$.

45. $r^4\csc^2\theta - R^4\csc\theta\cot\theta = r^4 \cdot \dfrac{1}{\sin^2\theta} - R^4 \cdot \dfrac{1}{\sin\theta} \cdot \dfrac{\cos\theta}{\sin\theta}$

$$= \frac{r^4}{\sin^2\theta} - \frac{R^4\cos\theta}{\sin^2\theta}$$

$$= \frac{r^4 - R^4\cos\theta}{\sin^2\theta}$$

Continued on next page.

This expression is zero only when the numerator is zero. Therefore,

$$r^4 - R^4\cos\theta = 0$$

$$-R^4\cos\theta = -r^4$$

$$\cos\theta = \frac{r^4}{R^4}$$

Therefore, when $\cos\theta = \dfrac{r^4}{R^4}$, then $r^4\csc^2\theta - R^4\csc\theta\cot\theta = 0$

49.
$$\cos^2\theta + \sin\theta = 0$$

$$1 - \sin^2\theta + \sin\theta = 0 \qquad\qquad \cos^2\theta = 1 - \sin^2\theta$$

$$-\sin^2\theta + \sin\theta + 1 = 0 \qquad\qquad \text{Rewrite in standard form}$$

$$\sin^2\theta - \sin\theta - 1 = 0 \qquad\qquad \text{Multiply both sides by } -1$$

We apply the quadratic formula with $a = 1$, $b = -1$, $c = -1$:

$$\sin\theta = \frac{-(-1) \pm \sqrt{(-1)^2 - 4(1)(-1)}}{2(1)}$$

$$= \frac{1 \pm \sqrt{5}}{2}$$

$$\sin\theta = \frac{1 + 2.236}{2} \qquad \text{or} \qquad \sin\theta = \frac{1 - 2.236}{2}$$

$$= 1.618 \qquad\qquad\qquad = -0.618$$

No Solution $\qquad\qquad \widehat{\theta} = 38.2°$ and θ is in QIII or QIV

$$\theta = 218.2° \text{ or } \theta = 321.8°$$

53. $\cos A = \sqrt{1 - \sin^2 A}$ $\qquad\qquad$ with A in QI

$$= \sqrt{1 - (\tfrac{2}{3})^2}$$

$$= \sqrt{1 - \frac{4}{9}}$$

$$= \sqrt{\frac{5}{9}}$$

$$= \frac{\sqrt{5}}{3}$$

$$\sin\frac{A}{2} = \sqrt{\frac{1 - \cos A}{2}}$$

$$= \sqrt{\frac{1 - \sqrt{5}/3}{2}}$$

$$= \sqrt{\frac{3 - \sqrt{5}}{6}}$$

57. $\tan \dfrac{A}{2} = \dfrac{\sin \frac{A}{2}}{\cos \frac{A}{2}}$

We know that $\cos A = \dfrac{\sqrt{5}}{3}$ and $\sin \dfrac{A}{2} = \sqrt{\dfrac{3 - \sqrt{5}}{6}}$ from Problem 53 above.

$$\cos \frac{A}{2} = \sqrt{\frac{1 + \cos A}{2}}$$

$$= \sqrt{\frac{1 + \sqrt{5}/3}{2}}$$

$$= \sqrt{\frac{3 + \sqrt{5}}{6}}$$

Therefore, $\tan \dfrac{A}{2} = \dfrac{\sqrt{\frac{3 - \sqrt{5}}{6}}}{\sqrt{\frac{3 + \sqrt{5}}{6}}}$

$$= \sqrt{\frac{3 - \sqrt{5}}{3 + \sqrt{5}}}$$

$$= \frac{3 - \sqrt{5}}{2} \qquad \text{Rationalize the denominator}$$

61. $\sin 22.5° = \sin \dfrac{1}{2}(45°)$

$$= \sqrt{\frac{1 - \cos 45°}{2}}$$

$$= \sqrt{\frac{1 - \sqrt{2}/2}{2}}$$

$$= \sqrt{\frac{2 - \sqrt{2}}{4}}$$

$$= \frac{\sqrt{2 - \sqrt{2}}}{2}$$

Problem Set 6.3

1. $\sin 2\theta = \dfrac{\sqrt{3}}{2}$

    $2\theta = 60° + 360°k \quad$ or $\quad 2\theta = 120° + 360°k$

    $\theta = 30° + 180°k \qquad \theta = 60° + 180°k$

    If we let $k = 0$ and 1, we get

    $\theta = 30° \qquad\qquad\qquad \theta = 60°$

    $\theta = 30° + 180° = 210° \quad \theta = 60° + 180° = 240°$

    Solutions: 30°, 60°, 210°, 240°

5. $\cos 3\theta = -1$

    $3\theta = 180° + 360°k$

    $\theta = 60° + 120°k$

    If we let $k = 0$, 1, and 2, we get

    $\theta = 60°$

    $\theta = 60° + 120° = 180°$

    $\theta = 60° + 240° = 300°$

    Solutions: 60°, 180°, 300°

9. $\sec 3x = -1$

    $\cos 3x = -1 \qquad\qquad \cos 3x = \dfrac{1}{\sec 3x}$

    $x = \dfrac{\pi}{3} + \dfrac{2k\pi}{3}$

    If we let $k = 0$, 1, and 2, we get

    $\theta = \dfrac{\pi}{3}$

    $\theta = \dfrac{\pi}{3} + \dfrac{2\pi}{3} = \pi$

    $\theta = \dfrac{\pi}{3} + \dfrac{4\pi}{3} = \dfrac{5\pi}{3}$

    Solutions: $\dfrac{\pi}{3}, \pi, \dfrac{5\pi}{3}$

13. $\sin 2\theta = \dfrac{1}{2}$

    $2\theta = 30° + 360°k$ or $\quad 2\theta = 150° + 360°k$

    $\theta = 15° + 180°k \qquad \theta = 75° + 180°k$

17. $\sin 10\theta = \dfrac{\sqrt{3}}{2}$

$10\theta = 60° + 360°k$ or $10\theta = 120° + 360°k$

$\theta = 6° + 36°k$ $\qquad\qquad$ $\theta = 12° + 36°k$

21. $\cos 2x \cos x - \sin 2x \sin x = -\dfrac{\sqrt{3}}{2}$

$\cos(2x + x) = -\dfrac{\sqrt{3}}{2}$ $\quad$ Sum formula

$\cos 3x = -\dfrac{\sqrt{3}}{2}$

$3x = \dfrac{5\pi}{6} + 2k\pi$ $\qquad$ or $\qquad$ $3x = \dfrac{7\pi}{6} + 2k\pi$

$x = \dfrac{5\pi}{18} + \dfrac{2k\pi}{3}$ $\qquad\qquad$ $x = \dfrac{7\pi}{18} + \dfrac{2k\pi}{3}$

$x = \dfrac{5\pi + 12k\pi}{18}$ $\qquad\qquad$ $x = \dfrac{7\pi + 12k\pi}{18}$

If we let $k = 0, 1,$ or 2, we get

$x = \dfrac{5\pi}{18}$ $\qquad\qquad\qquad$ $x = \dfrac{7\pi}{18}$

$x = \dfrac{5\pi + 12\pi}{18} = \dfrac{17\pi}{18}$ $\qquad$ $x = \dfrac{7\pi + 12\pi}{18} = \dfrac{19\pi}{18}$

$x = \dfrac{5\pi + 24\pi}{18} = \dfrac{29\pi}{18}$ $\qquad$ $x = \dfrac{7\pi + 24\pi}{18} = \dfrac{31\pi}{18}$

Solutions: $\dfrac{5\pi}{18}, \dfrac{7\pi}{18}, \dfrac{17\pi}{18}, \dfrac{19\pi}{18}, \dfrac{29\pi}{18}, \dfrac{31\pi}{18}$

25. $\sin^2 4x = 1$

$\sin 4x = \pm 1$

$4x = \dfrac{\pi}{2} + 2k\pi$ $\quad$ or $4x = \dfrac{3\pi}{2} + 2k\pi$

$x = \dfrac{\pi}{8} + \dfrac{k\pi}{2}$ $\qquad$ $x = \dfrac{3\pi}{8} + \dfrac{k\pi}{2}$

(We could also write this as $\quad 4x = \dfrac{\pi}{2} + k\pi$ $\quad$ or $x = \dfrac{\pi}{8} + \dfrac{k\pi}{4}$)

29. $2\sin^2 3\theta + \sin 3\theta - 1 = 0$

$(2\sin 3\theta - 1)(\sin 3\theta + 1) = 0$

$2\sin 3\theta - = 0 \qquad\text{or}\qquad \sin 3\theta + 1 = 0$

$2\sin 3\theta = 1 \qquad\qquad\qquad \sin 3\theta = -1$

$\sin 3\theta = \tfrac{1}{2} \qquad\qquad\qquad\qquad 3\theta = 270° + 360°\,k$

$3\theta = 30° + 360°k \ \text{ or } \ 3\theta = 150° + 360°k \qquad \theta = 90° + 120°\,k$

$\theta = 10° + 120°k \qquad \theta = 50° + 120°k$

33. $\tan^2 3\theta = 3$

$\tan 3\theta = \pm\sqrt{3}$

$3\theta = 60° + 180°k \ \text{ or } \qquad 3\theta = 120° + 180°k$

$\theta = 20° + 60°k \qquad\qquad\quad \theta = 40° + 60°k$

37. $\sin\theta + \cos\theta = -1$

$\sin^2\theta + 2\sin\theta\cos\theta + \cos^2\theta = 1 \qquad\qquad$ Square both sides

$1 + 2\sin\theta\cos\theta = 1 \qquad\qquad\qquad \sin^2\theta + \cos^2\theta = 1$

$\sin 2\theta = 0 \qquad\qquad\qquad\qquad \sin 2\theta = 2\sin\theta\cos\theta$

$2\theta = 0° + 360°k \quad\text{or}\quad 2\theta = 180° + 360°k$

$\theta = 180°k \qquad\qquad\qquad \theta = 90° + 180°k$

If we let $k = 0$ and 1, we get

$\theta = 0° \qquad \theta = 90° \qquad\qquad\qquad$ Possible solutions

$\theta = 180° \qquad \theta = 90° + 180° = 270° \qquad$ Possible solutions

Since we squared both sides, we must check:

$\sin 0° + \cos 0° \overset{?}{=} -1 \qquad\qquad \sin 90° + \cos 90° \overset{?}{=} -1$

$0 + 1 \overset{?}{=} -1 \qquad\qquad\qquad 1 + 0 \overset{?}{=} -1$

$1 \neq -1 \qquad\qquad\qquad\qquad 1 \neq -1$

$\sin 180° + \cos 180° \overset{?}{=} -1 \qquad\qquad \sin 270° + \cos 270° \overset{?}{=} -1$

$0 + (-1) \overset{?}{=} -1 \qquad\qquad\qquad -1 + 0 \overset{?}{=} -1$

$-1 = -1 \qquad\qquad\qquad\qquad -1 = -1$

The solutions are 180° and 270°.

41. $4\cos^2 3\theta - 8\cos 3\theta + 1 = 0$

$$\cos 3\theta = \frac{-(-8) \pm \sqrt{(-8)^2 - 4(4)(1)}}{2(4)} \qquad\qquad a = 4, b = -8, c = 1$$

$$= \frac{8 \pm \sqrt{48}}{8}$$

$$\cos 3\theta = \frac{8 \pm 6.9282}{8}$$

$$\cos 3\theta = 1.8660 \quad \text{or} \quad \cos 3\theta = 0.1340$$

No solution $\qquad\qquad 3\theta = 82.3 + 360°k \quad 3\theta = 277.7° + 360°$

$$\theta = 27.4° + 120°k \quad \theta = 92.6° + 120°k$$

If we let $k = 0, 1$ and 2, we get

$$\theta = 27.4° \qquad\qquad \theta = 92.6°$$

$$\theta = 147.4° \qquad\qquad \theta = 212.6°$$

$$\theta = 267.4° \qquad\qquad \theta = 332.6°$$

45. We want to find t when $h = 100$.

$$100 = 139 - 125\cos\frac{\pi}{10}t$$

$$125\cos\frac{\pi}{10t} = 39$$

$$\cos\frac{\pi}{10}t = 0.312$$

$$\frac{\pi}{10}t = 1.253 + 2\pi k \quad \text{or} \quad \frac{\pi}{10}t = (2\pi - 1.253) + 2\pi k$$

$$t = 4.0 + 20k \qquad\qquad \frac{\pi}{10}t = 5.030 + 2\pi k$$

$$t = 16.0 + 20k$$

It will be at 100 ft after 4.0 min, 16.0 min, 24.0 min, 28.0 min, and so on.

49. $d = 10\tan\pi t$ where $d = 10$

$$10 = 10\tan\pi t$$

$$\tan\pi t = 1$$

$$\pi t = \frac{\pi}{4} + k\pi$$

$$t = \frac{1}{4} + k$$

$$t = \frac{1}{4} \text{ second and every second after that}$$

53. $\dfrac{\sin x}{1 + \cos x} = \dfrac{\sin x}{1 + \cos x} \cdot \dfrac{1 - \cos x}{1 - \cos x}$ Multiply by a fraction equal to one

$$= \dfrac{\sin x(1 - \cos x)}{1 - \cos^2 x}$$ Multiply

$$= \dfrac{\sin x(1 - \cos x)}{\sin^2 x}$$ Pythagorean identity

$$= \dfrac{1 - \cos x}{\sin x}$$ Reduce

57. $\tan \dfrac{A}{2} = \dfrac{\sin \frac{A}{2}}{\cos \frac{A}{2}}$ Ratio identity

$$= \dfrac{\sqrt{\frac{1 - \cos A}{2}}}{\sqrt{\frac{1 + \cos A}{2}}}$$ Half-angle formulas

$$= \dfrac{\sqrt{1 - \cos A}}{\sqrt{1 + \cos A}}$$ Divide

$$= \dfrac{\sqrt{1 - \cos A}}{\sqrt{1 + \cos A}} \cdot \dfrac{\sqrt{1 + \cos A}}{\sqrt{1 + \cos A}}$$ Multiply by fraction equal to one

$$= \dfrac{\sqrt{1 - \cos^2 A}}{1 + \cos A}$$ Multiply

$$= \dfrac{\sqrt{\sin^2 A}}{1 + \cos A}$$ Pythagorean identity

$$= \dfrac{\sin A}{1 + \cos A}$$ Simplify

61. If $90° \leq A \leq 180°$, then $45° \leq \dfrac{A}{2} \leq 90°$.

Also, if $\sin a = \dfrac{1}{3}$ and A is in QII, then

$$\cos A = -\sqrt{1 - \sin^2 A}$$

$$= -\sqrt{1 - \dfrac{1}{9}} = -\sqrt{\dfrac{8}{9}} = -\dfrac{2\sqrt{2}}{3}$$

Therefore, $\cos \dfrac{A}{2} = \sqrt{\dfrac{1 + \cos A}{2}}$

$$= \sqrt{\frac{1 + \left(-\frac{2\sqrt{2}}{3}\right)}{2}}$$

$$= \sqrt{\frac{3 - 2\sqrt{2}}{6}}$$

65. $\csc(A + B) = \dfrac{1}{\sin(A + B)}$

$$= \frac{1}{\sin A \cos B + \cos A \sin B}$$

If $\sin B = \dfrac{3}{5}$ with B in QI, then

$$\cos B = \sqrt{1 - \sin^2 B}$$

$$= \sqrt{1 - \frac{9}{25}} = \sqrt{\frac{16}{25}} = \frac{4}{5}$$

From Problem 61, we have $\cos A = -\dfrac{2\sqrt{2}}{3}$.

Substituting these above, we get:

$$\csc(A + B) = \frac{1}{\frac{1}{3}\left(\frac{4}{5}\right) + \left(-\frac{2\sqrt{2}}{3}\right)\left(\frac{3}{5}\right)}$$

$$= \frac{1}{\frac{4}{15} - \frac{6\sqrt{2}}{15}}$$

$$= \frac{1}{\frac{4 - 6\sqrt{2}}{15}}$$

$$= \frac{15}{4 - 6\sqrt{2}}$$

Problem Set 6.4

1. $\sin t = x$ and $\cos t = y$

$\sin^2 t + \cos^2 t = 1$	Pythagorean identity
$x^2 + y^2 = 1$	Substitute given values

The graph is a circle with its center at the origin and $r = 1$.

5. $2\sin t = x$ and $4\cos t = y$

$\qquad \sin t = \dfrac{x}{2} \qquad\qquad \cos t = \dfrac{y}{4}$

$\qquad \sin^2 t + \cos^2 t = 1$

$\qquad (\dfrac{x}{2})^2 + (\dfrac{y}{4})^2 = 1$

$\qquad \dfrac{x^2}{4} + \dfrac{y^2}{16} = 1$

The graph is an ellipse with center at the origin. The intercepts on the major axis are $(0,4)$ and $(0,-4)$. The intercepts on the minor axis are $(2,0)$ and $(-2,0)$.

9. $\sin t - 2 = x$ and $\cos t - 3 = y$

$\qquad \sin t = x + 2 \qquad\qquad \cos t = y + 3$

$\qquad \sin^2 t + \cos^2 t = 1$

$\qquad (x+2)^2 + (y+3)^2 = 1$

The graph is a circle with center at $(-2,-3)$ and $r = 1$.

13. $3\cos t - 3 = x$ and $3\sin t + 1 = y$

$\qquad 3\cos t = x + 3 \qquad\qquad 3\sin t = y - 1$

$\qquad \cos t = \dfrac{x+3}{3} \qquad\qquad \sin t = \dfrac{y-1}{3}$

$\qquad \cos^2 t + \sin^2 t = 1$

$\qquad \left(\dfrac{x+3}{3}\right)^2 + \left(\dfrac{y-1}{3}\right)^2 = 1$

$\qquad \dfrac{(x+3)^2}{9} + \dfrac{(y-1)^2}{9} = 1$

$\qquad (x+3)^2 + (y-1)^2 = 9$

The graph is a circle with center at $(-3,1)$ and $r = 3$.

17. $3\sec t = x$ and $3\tan t = y$

$\qquad \sec t = \dfrac{x}{3} \qquad\qquad \tan t = \dfrac{y}{3}$

$\qquad \tan^2 t + 1 = \sec^2 t$

$\qquad \left(\dfrac{y}{3}\right)^2 + 1 = \left(\dfrac{x}{3}\right)^2$

$\qquad 1 = \dfrac{x^2}{9} - \dfrac{y^2}{9}$ or $\dfrac{x^2}{9} - \dfrac{y^2}{9} = 1$

21.
$$\cos 2t = x \quad \text{and} \quad \sin t = y$$
$$2\cos^2 t - 1 = x$$
$$2\cos^2 t = x + 1$$
$$\cos^2 t = \frac{x+1}{2}$$

$\cos^2 t + \sin^2 t = 1$ Pythagorean identity

$\dfrac{x+1}{2} + y^2 = 1$ Substitute values from above

$x + 1 + 2y^2 = 2$ Multiply both sides by 2

$x = 1 - 2y^2$ Solve for x

25. $3\sin t = x \quad \text{and} \quad 2\sin t = y$

     $\sin t = \dfrac{x}{3} \qquad\qquad \sin t = \dfrac{y}{2}$

     $\sin t = \sin t$ Reflexive property of equality

     $\dfrac{x}{3} = \dfrac{y}{2}$ Substitute values from above

     $2x = 3y$ Multiply both sides by 6

29. Let $\alpha = \cos^{-1}\dfrac{1}{2}$. Then $\cos\alpha = \dfrac{1}{2}$ and $0 \le \alpha \le \pi$.

Therefore, $\alpha = \pi/3$. We want to find $\sin\alpha = \sin\dfrac{\pi}{3} = \dfrac{\sqrt{3}}{2}$

33. Let $\alpha = \sin^{-1} x$. Then $\sin\alpha = \dfrac{x}{1}$ and $-\dfrac{\pi}{2} \le \alpha \le \dfrac{\pi}{2}$.

We draw a triangle and we find the adjacent side

using the Pythagorean theorem

Therefore, $\cos\alpha = \dfrac{\sqrt{1-x^2}}{1} = \sqrt{1-x^2}$.

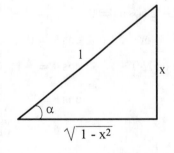

37. $8\sin 3x \cos 2x = 8 \cdot \dfrac{1}{2}[\sin(3x+2x) + \sin(3x - 2x)]$

                 $= 4(\sin 5x + \sin x)$

Chapter 6 Test

1. $2\sin\theta - 1 = 0$

  $2\sin\theta = 1$ Add 1 to both sides

  $\sin\theta = \dfrac{1}{2}$ Divide both sides by 2

  $\theta = 30°$ or $150°$ $\hat{\theta} = 30°$ and θ is in QI or QII

2. $\sqrt{3}\tan\theta + 1 = 0$

  $\sqrt{3}\tan\theta = -1$ Subtract 1 from both sides

  $\tan\theta = -\dfrac{1}{\sqrt{3}}$ Divide both sides by $\sqrt{3}$

  $\theta = 150°$ or $330°$ $\hat{\theta} = 30°$ and θ is in QII and QIV

3. $\cos\theta - 2\sin\theta\cos\theta = 0$

  $\cos\theta(1 - 2\sin\theta) = 0$ Factor

  $\cos\theta = 0$ or $1 - 2\sin\theta = 0$ Set each factor $= 0$

  $\theta = 90°$ or $270°$ $-2\sin\theta = -1$ Solve each equation

  $\sin\theta = \dfrac{1}{2}$

  $\theta = 30°$ or $150°$

4. $\tan\theta - 2\cos\theta\tan\theta = 0$

  $\tan\theta(1 - 2\cos\theta) = 0$ Factor

  $\tan\theta = 0$ or $1 - 2\cos\theta = 0$ Set each factor $= 0$

  $\theta = 0$ or $180°$ $-2\cos\theta = -1$ Solve each equation

  $\cos\theta = \dfrac{1}{2}$

  $\theta = 60°$ or $300°$

5. $4\cos\theta - 2\sec\theta = 0$

  $4\cos\theta - \dfrac{2}{\cos\theta} = 0$ $\sec\theta = \dfrac{1}{\cos\theta}$

  $4\cos^2\theta - 2 = 0$ Multiply both sides by $\cos\theta$

  $4\cos^2\theta = 2$ Add 2 to both sides

$$\cos^2 \theta = \tfrac{1}{2} \qquad \text{Divide both sides by 4}$$

$$\cos \theta = \pm \frac{1}{\sqrt{2}} \qquad \text{Take square root of both sides}$$

$$\theta = 45°, 135°, 225°, 315° \qquad \widehat{\theta} = 45°$$

6.
$$2\sin\theta - \csc\theta = 1$$

$$2\sin\theta - \frac{1}{\sin\theta} = 1 \qquad \csc\theta = \frac{1}{\sin\theta}$$

$$2\sin^2\theta - 1 = \sin\theta \qquad \text{Multiply both sides by } \sin\theta$$

$$2\sin^2\theta - \sin\theta - 1 = 0 \qquad \text{Rewrite in standard form}$$

$$(2\sin\theta + 1)(\sin\theta - 1) = 0 \qquad \text{Factor}$$

$$2\sin\theta + 1 = 0 \quad \text{or} \quad \sin\theta - 1 = 0 \qquad \text{Set each factor} = 0$$

$$2\sin\theta = -1 \qquad\qquad \sin\theta = 1 \qquad \text{Solve each equation}$$

$$\sin\theta = -\frac{1}{2} \qquad\qquad \theta = 90°$$

$$\theta = 210° \text{ or } 330°$$

7.
$$\sin\frac{\theta}{2} + \cos\theta = 0$$

$$\pm\sqrt{\frac{1-\cos\theta}{2}} + \cos\theta = 0 \qquad \text{Half-angle formula}$$

$$\pm\sqrt{\frac{1-\cos\theta}{2}} = -\cos\theta \qquad \text{Subtract } \cos\theta \text{ from both sides}$$

$$\frac{1-\cos\theta}{2} = \cos^2\theta \qquad \text{Square both sides}$$

$$1 - \cos\theta = 2\cos^2\theta \qquad \text{Multiply both sides by 2}$$

$$2\cos^2\theta + \cos\theta - 1 = 0 \qquad \text{Rewrite in standard form}$$

$$(2\cos\theta - 1)(\cos\theta + 1) = 0 \qquad \text{Factor}$$

$$2\cos\theta - 1 = 0 \quad \text{or} \quad \cos\theta + 1 = 0 \qquad \text{Set each factor} = 0$$

$$2\cos\theta = 1 \qquad\qquad \cos\theta = -1 \qquad \text{Solve each equation and check}$$

$$\cos\theta = 1/2 \qquad\qquad \theta = 180°$$

$$\theta = 60° \text{ or } 300°$$

We check each possible solution on the next page.

Check:

$$\sin \frac{60°}{2} + \cos 60° \overset{?}{=} 0 \qquad\qquad \sin \frac{180°}{2} + \cos 180° \overset{?}{=} 0$$

$$\sin 30° + \cos 60° \overset{?}{=} 0 \qquad\qquad \sin 90° + \cos 180° \overset{?}{=} 0$$

$$\frac{1}{2} + \frac{1}{2} \overset{?}{=} 0 \qquad\qquad\qquad 1 + (-1) \overset{?}{=} 0$$

$$1 \neq 0 \qquad\qquad\qquad\qquad 0 = 0$$

$$\sin \frac{300°}{2} + \cos 300° \overset{?}{=} 0$$

$$\sin 150° + \cos 300° \overset{?}{=} 0$$

$$\frac{1}{2} + \frac{1}{2} \overset{?}{=} 0$$

$$1 \neq 0$$

The only answer that checks is $\theta = 180°$.

8.
$$\cos \frac{\theta}{2} - \cos \theta = 0$$

$\pm \sqrt{\dfrac{1 + \cos \theta}{2}} - \cos \theta = 0$	Half-angle formula
$\pm \sqrt{\dfrac{1 + \cos \theta}{2}} = \cos \theta$	Add $\cos \theta$ to both sides
$\dfrac{1 + \cos \theta}{2} = \cos^2 \theta$	Square both sides
$1 + \cos \theta = 2\cos^2 \theta$	Multiply both sides by 2
$2\cos^2 \theta - \cos \theta - 1 = 0$	Rewrite in standard form
$(2\cos \theta + 1)(\cos \theta - 1) = 0$	Factor
$2\cos \theta + 1 = 0 \quad \text{or} \quad \cos \theta - 1 = 0$	Set each factor $= 0$
$2\cos \theta = -1 \qquad\qquad \cos \theta = 1$	Solve each equation and check
$\cos \theta = -\dfrac{1}{2} \qquad\qquad \theta = 0°$	
$\theta = 120° \text{ or } 240°$	

Check:

$$\cos \frac{0°}{2} - \cos 0° \overset{?}{=} 0 \qquad\qquad \cos \frac{120°}{2} - \cos 120° \overset{?}{=} 0$$

$$\cos 0° - \cos 0° \overset{?}{=} 0 \qquad\qquad \cos 60° - \cos 120° \overset{?}{=} 0$$

$$0 = 0 \qquad\qquad\qquad \frac{1}{2} - \left(-\frac{1}{2}\right) \overset{?}{=} 0$$

$$1 \neq 0$$

$$\cos \frac{240°}{2} - \cos 240° \overset{?}{=} 0$$

$$\cos 120° - \cos 240° \overset{?}{=} 0$$

$$\frac{1}{2} - \frac{1}{2} \overset{?}{=} 0$$

$$0 = 0$$

The solutions are $0°$ and $240°$.

9.

$4\cos 2\theta + 2\sin\theta = 1$	
$4(1 - 2\sin^2\theta) + 2\sin\theta = 1$	$\cos 2\theta = 1 - 2\sin^2\theta$
$4 - 8\sin^2\theta + 2\sin\theta = 1$	Simplify
$-8\sin^2\theta + 2\sin\theta + 3 = 0$	Subtract 1 from both sides
$8\sin^2\theta - 2\sin\theta - 3 = 0$	Multiply both sides by -1
$(4\sin\theta - 3)(2\sin\theta + 1) = 0$	Factor
$4\sin\theta - 3 = 0 \qquad\text{or}\qquad 2\sin\theta + 1 = 0$	Set each factor $= 0$
$4\sin\theta = 3 \qquad\qquad 2\sin\theta = -1$	Solve each equation
$\sin\theta = \dfrac{3}{4} \qquad\qquad \sin\theta = -\dfrac{1}{2}$	
$\theta = 48.6° \text{ or } 131.4° \qquad \theta = 210° \text{ or } 330°$	

10. $\sin(3\theta - 45°) = -\dfrac{\sqrt{3}}{2}$ (The reference angle is $60°$)

$$3\theta - 45° = 240° + 360°k \quad\text{or}\quad 3\theta - 45° = 300° + 360°k$$

$$3\theta = 285° + 360°k \qquad\qquad 3\theta = 345° + 360°k$$

$$\theta = 95° + 120°k \qquad\qquad \theta = 115° + 120°k$$

Let $k = 0, 1$ and 2, then

$$\theta = 95° \qquad\text{or}\qquad \theta = 115°$$

$$\theta = 215° \qquad\qquad \theta = 235°$$

$$\theta = 335° \qquad\qquad \theta = 355°$$

11.
$$\sin \theta + \cos \theta = 1$$

$\sin \theta = 1 - \cos \theta$	Subtract $\cos \theta$ from both sides
$\sin^2\theta = 1 - 2\cos\theta + \cos^2\theta$	Square both sides
$1 - \cos^2\theta = 1 - 2\cos\theta + \cos^2\theta$	Pythagorean identity
$2\cos^2\theta - 2\cos\theta = 0$	Rewrite in standard form
$2\cos\theta(\cos\theta - 1) = 0$	Factor
$\cos\theta = 0 \quad$ or $\quad \cos\theta - 1 = 0$	Set each factor $= 0$
$\theta = 90° \text{ or } 270° \qquad \cos\theta = 1$	Solve each equation and check
$\theta = 0°$	

Check:

$$\sin 0° + \cos 0° \overset{?}{=} 0 \qquad\qquad \sin 90° + \cos 90° \overset{?}{=} 1$$

$$0 + 1 \overset{?}{=} 1 \qquad\qquad\qquad 1 + 0 \overset{?}{=} 1$$

$$1 = 1 \qquad\qquad\qquad\qquad 1 = 1$$

$$\sin 270° + \cos 270° \overset{?}{=} 1$$

$$-1 + 0 \overset{?}{=} 1$$

$$-1 \neq 1$$

The solutions are $\theta = 0°$ and $90°$.

12.
$$\sin \theta - \cos \theta = 1$$

$\sin \theta = 1 + \cos \theta$	Add $\cos \theta$ to both sides
$\sin^2\theta = 1 + 2\cos\theta + \cos^2\theta$	Square both sides
$1 - \cos^2\theta = 1 + 2\cos\theta + \cos^2\theta$	Pythagorean identity
$2\cos^2\theta + 2\cos\theta = 0$	Rewrite in standard form
$2\cos\theta(\cos\theta + 1) = 0$	Factor
$2\cos\theta = 0 \quad$ or $\quad \cos\theta + 1 = 0$	Set each factor $= 0$
$\cos\theta = 0 \qquad\qquad \cos\theta = -1$	Solve each equation and check
$\theta = 90° \text{ or } 270° \qquad \theta = 180°$	

Check:

$$\sin 90° - \cos 90° \overset{?}{=} 1 \qquad\qquad \sin 180° - \cos 180° \overset{?}{=} 1$$

$$1 - 0 \overset{?}{=} 1 \qquad\qquad\qquad 0 - (-1) \overset{?}{=} 1$$

$$1 = 1 \qquad\qquad\qquad\qquad 1 = 1$$

$$\sin 270° - \cos 270° \overset{?}{=} 1$$

$$-1 - 0 \overset{?}{=} 1$$

$$-1 \neq 1$$

The solutions are $\theta = 90°$ or $180°$.

13.
$$\cos 2x - 3\cos x = -2$$

$2\cos^2 x - 1 - 3\cos x = -2$	$\cos 2x = 2\cos^2 x - 1$
$2\cos^2 x - 3\cos x + 1 = 0$	Add 2 to both sides
$(2\cos x - 1)(\cos x - 1) = 0$	Factor
$2\cos x - 1 = 0 \quad$ or $\quad \cos x - 1 = 0$	Set each factor $= 0$
$2\cos x = 1 \qquad\qquad \cos x = 1$	Solve each equation

$$x = 0 + 2k\pi$$

$$\cos x = \frac{1}{2} \qquad\qquad x = 2k\pi$$

$$x = \frac{\pi}{3} + 2k\pi \text{ or}$$

$$x = \frac{5\pi}{3} + 2k\pi$$

14. $\sqrt{3}\sin x - \cos x = 0$

$\sqrt{3}\sin x = \cos x$	Add $\cos x$ to both sides
$\sqrt{3}\,\dfrac{\sin x}{\cos x} = 1$	Divide both sides by $\cos x$
$\sqrt{3}\tan x = 1$	Ratio identity
$\tan x = \dfrac{1}{\sqrt{3}}$	Divide both sides by $\sqrt{3}$.
$x = \dfrac{\pi}{6} + k\pi$	(This answer can also be written as $\dfrac{\pi}{6} + 2k\pi$ or $\dfrac{7\pi}{6} + 2k\pi$.)

15. $\sin 2x \cos x + \cos 2x \sin x = -1$

$\sin(2x + x) = -1$	Sum formula
$\sin 3x = -1$	Simplify

$$3x = \frac{3\pi}{2} + 2k\pi$$

$$x = \frac{\pi}{2} + \frac{2k\pi}{3}$$

16. $\sin^3 4x = 1$

$\quad\quad \sin 4x = 1$ $\quad\quad\quad\quad\quad\quad$ Take cube root of both sides

$\quad\quad\quad 4x = \dfrac{\pi}{2} + 2k\pi$

$\quad\quad\quad\quad x = \dfrac{\pi}{8} + \dfrac{\pi}{2}$

17. $\quad\quad 5\sin^2\theta - 3\sin\theta = 2$

$\quad\quad 5\sin^2\theta - 3\sin\theta - 2 = 0$ $\quad\quad\quad\quad$ Subtract 2 from both sides

$\quad\quad (5\sin\theta + 2)(\sin\theta - 1) = 0$ $\quad\quad\quad$ Factor

$\quad\quad 5\sin\theta + 2 = 0 \quad\quad \text{or} \quad\quad \sin\theta - 1 = 0$ $\quad\quad$ Set each factor $= 0$

$\quad\quad\quad\quad 5\sin\theta = -2 \quad\quad\quad\quad\quad \sin\theta = 1$ $\quad\quad\quad$ Solve each equation

$\quad\quad\quad\quad\quad \sin\theta = -\dfrac{2}{5} \quad\quad\quad\quad\quad \theta = 90°$

$\quad\quad\quad\quad\quad \sin\theta = -0.4$

$\quad\quad\quad\quad\quad \widehat{\theta} = 23.6°$

$\quad\quad\quad\quad\quad \theta = 203.6° \text{ or } 336.4°$

18. $\quad\quad 4\cos^2\theta - 4\cos\theta = 2$

$\quad\quad 4\cos^2\theta - 4\cos\theta - 2 = 0$ $\quad\quad\quad\quad$ Subtract 2 from both sides

$\quad\quad 2\cos^2\theta - 2\cos\theta - 1 = 0$ $\quad\quad\quad\quad$ Divide both sides by 2

$\quad\quad \cos\theta = \dfrac{-(-2) \pm \sqrt{(-2)^2 - 4(2)(-1)}}{2(2)}$ $\quad\quad a = 2, b = -2, c = -1$

$\quad\quad\quad\quad = \dfrac{2 \pm \sqrt{12}}{4}$

$\quad\quad \cos\theta = \dfrac{2 \pm 3.4641}{4}$

$\quad\quad \cos\theta = 1.3660 \quad\quad \text{or} \quad\quad \cos\theta = -0.3660$

$\quad\quad \text{No solution} \quad\quad\quad\quad\quad\quad \theta = 111.5° \text{ or } 248.5°$

19. $\quad 3\cos t = x \quad\quad \text{and} \quad\quad 3\sin t = y$

$\quad\quad\quad \cos t = \dfrac{x}{3} \quad\quad\quad\quad\quad\quad \sin t = \dfrac{y}{3}$

$\quad \cos^2 t + \sin^2 t = 1$

$\quad \left(\dfrac{x}{3}\right)^2 + \left(\dfrac{y}{3}\right)^2 = 1$

$$\frac{x^2}{9} + \frac{y^2}{9} = 1$$

$$x^2 + y^2 = 9$$

The graph is a circle with a center at $(0, 0)$ and $r = 3$.

20. $\quad \sec t = x \qquad$ and $\qquad \tan t = y$

$$\sec^2 t = \tan^2 t + 1$$

$$x^2 = y^2 + 1$$

$$x^2 - y^2 = 1$$

The graph is a hyperbola with center at $(0, 0)$, vertices at $(\pm 1, 0)$, and asymptotes of $y = x$ and $y = -x$.

21. $\quad 3 + 2 \sin t = x \qquad\qquad 1 + 2 \cos t = y$

$$2 \sin t = x - 3 \qquad\qquad 2 \cos t = y - 1$$

$$\sin t = \frac{x - 3}{2} \qquad\qquad \cos t = \frac{y - 1}{2}$$

$$\sin^2 t + \cos^2 t = 1$$

$$\left(\frac{x - 3}{2}\right)^2 + \left(\frac{y - 1}{2}\right)^2 = 1$$

$$\frac{(x - 3)^2}{4} + \frac{(y - 1)^2}{4} = 1$$

$$(x - 3)^2 + (y - 1)^2 = 4$$

The graph is a circle with center at $(3, 1)$ and $r = 2$.

22. $\quad 3 \cos t - 3 = x \qquad$ and $\qquad 3 \sin t + 1 = y$

$$3 \cos t = x + 3 \qquad\qquad 3 \sin t = y - 1$$

$$\cos t = \frac{x + 3}{3} \qquad\qquad \sin t = \frac{y - 1}{3}$$

$$\cos^2 t + \sin^2 t = 1$$

$$\left(\frac{x + 3}{3}\right)^2 + \left(\frac{y - 1}{3}\right)^2 = 1$$

$$\frac{(x + 3)^2}{9} + \frac{(y - 1)^2}{9} = 1$$

$$(x + 3)^2 + (y - 1)^2 = 9$$

The graph is a circle with center at $(-3, 1)$ and $r = 3$.

CHAPTER 7 Triangles

Problem Set 7.1

1. $$\frac{b}{\sin B} = \frac{a}{\sin A}$$ Law of Sines

$$\frac{b}{\sin 60°} = \frac{12}{\sin 40°}$$ Substitute given values

$$b = \frac{12 \sin 60°}{\sin 40°}$$ Multiply both sides by $\sin 60°$

$$= \frac{12(0.8660)}{0.6428}$$ Calculator

$$= 16 \text{ cm.}$$ Round to 2 significant digits

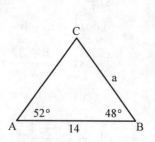

5. $$\frac{c}{\sin C} = \frac{a}{\sin A}$$ Law of Sines

$$\frac{c}{\sin 100°} = \frac{24}{\sin 10°}$$ Substitute given values

$$c = \frac{24 \sin 100°}{\sin 10°}$$ Multiply both sides by $\sin 100°$

$$= \frac{24(0.9848)}{0.1736}$$ Calculator

$$= 140 \text{ yards}$$ Round to 2 significant digits

9. $$C = 180° - (A + B)$$
$$= 180° - (52° + 48°)$$
$$= 180° - 100°$$
$$= 80°$$

$$\frac{a}{\sin A} = \frac{c}{\sin C}$$ Law of Sines

$$\frac{a}{\sin 52°} = \frac{14}{\sin 80°}$$ Substitute given values

$$a = \frac{14 \sin 52°}{\sin 80°}$$ Multiply both sides by $\sin 52°$

$$= \frac{14(0.7880)}{0.9848}$$ Calculator

$$= 11 \text{ cm}$$ Round to 2 significant digits

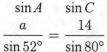

13.
$$C = 180° − (A + B)$$
$$= 180° − (46° + 95°)$$
$$= 180° − 141°$$
$$= 96°$$

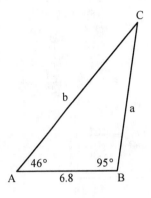

$$\frac{a}{\sin A} = \frac{c}{\sin C} \qquad \text{Law of Sines}$$

$$\frac{a}{\sin 46°} = \frac{6.8}{\sin 39°} \qquad \text{Substitute given values}$$

$$a = \frac{6.8 \sin 46°}{\sin 39°} \qquad \text{Multiply both sides by } \sin 46°$$

$$= \frac{6.8(0.7193)}{0.6293} \qquad \text{Calculator}$$

$$= 7.8 \text{ meters} \qquad \text{Round to 2 significant digits}$$

$$\frac{c}{\sin C} = \frac{b}{\sin B} \qquad \text{Law of Sines}$$

$$\frac{6.8}{\sin 39°} = \frac{b}{\sin 95°} \qquad \text{Substitute given values}$$

$$b = \frac{6.8 \sin 95°}{\sin 39°} \qquad \text{Multiply both sides by } \sin 95°$$

$$= \frac{6.8(0.9962)}{0.6293} \qquad \text{Calculator}$$

$$= 11 \text{ meters} \qquad \text{Round to 2 significant digits}$$

17. $A = 180° − (B + C)$
$$= 180° − (13.4° + 24.8°)$$
$$= 180° − 38.2°$$
$$= 141.8°$$

Continued on next page.

$$\frac{a}{\sin A} = \frac{b}{\sin B}$$ Law of Sines

$$\frac{315}{\sin 141.8°} = \frac{b}{\sin 13.4°}$$ Substitute given values

$$a = \frac{315 \sin 13.4°}{\sin 141.8°}$$ Multiply both sides by $\sin 13.4°$

$$= \frac{315(0.2317)}{0.6184}$$ Calculator

$$= 118 \text{ centimeters}$$ Round to 3 significant digits

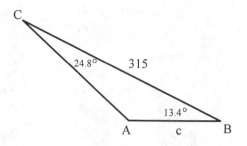

$$\frac{c}{\sin C} = \frac{a}{\sin A}$$ Law of Sines

$$\frac{c}{\sin 24.8°} = \frac{315}{\sin 141.8°}$$ Substitute given values

$$c = \frac{315(\sin 24.8°)}{\sin 141.8°}$$ Multiply both sides by $\sin 24.8°$

$$= \frac{315(0.4195)}{0.6184}$$ Calculator

$$= 214 \text{ centimeters}$$ Round to 3 significant digits

21. $S = r\theta$ (θ is $\angle C$)   Arc length formula

$11 = 12 \cdot \theta$ Substitute given values

$\theta = \dfrac{11}{12}$ Divide both sides by 12

We have $\angle C = \dfrac{11}{12}$ radians. Converting this to degrees, we get:

$$\angle C = \left(\frac{11}{12} \cdot \frac{180}{\pi} \right)^{\circ} = 53°$$

$$D = 180° - (C + A)$$
$$= 180° - (53° + 31°)$$
$$= 180° - 84°$$
$$= 96°$$

Using the Law of Sines, we get

$$\frac{x + r}{\sin D} = \frac{r}{\sin A}$$

$$\frac{x + 12}{\sin 96°} = \frac{12}{\sin 31°}$$

$$x + 12 = \frac{12(\sin 96°)}{\sin 31°}$$

$$x + 12 = \frac{12(0.9945)}{0.5150}$$

$$x + 12 = 23$$

$$x = 11$$

25. We find the missing angles first:

$$\angle ABD = 180° - 64° = 116°$$

$$\angle ADB = 180° - (46° + 116°) = 180° - 162° = 18°$$

Now we find BD using the Law of Sines

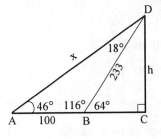

$$\frac{BD}{\sin A} = \frac{AB}{\sin ADB}$$

$$\frac{BD}{\sin 46°} = \frac{100}{\sin 18°}$$

$$BD = \frac{100 \sin 46°}{\sin 18°}$$

$$= \frac{100(0.7193)}{0.3090}$$

$$= 233$$

Then we find h, using the sine ratio:

$$\sin 64° = \frac{h}{233}$$

$$h = 233 \sin 64°$$

$$= 233(0.8988)$$

$$= 209 \text{ feet}$$

29. We find the missing angles first:

$$\angle ADB = 90° - 35° = 55°$$

$$\angle ADC = 180° - 55° = 125°$$

$$\angle C = 180° - (125° + 0.5°) = 180° - 125.5° = 54.5°$$

Continued on next page.

Now we find AD using the Law of Sines

$$\frac{AD}{\sin 54.5°} = \frac{110}{\sin 0.5°}$$

$$AD = \frac{110 \sin 54.5°}{\sin 0.5°}$$

$$BD = \frac{110(0.8141)}{0.0087}$$

$$= 10,262$$

Then we find h, using the sine ratio:

$$\sin 35° = \frac{h}{10,262}$$

$$h = 10,262 \sin 35°$$

$$= 10,262(0.5736)$$

$$= 5,900 \text{ feet}$$

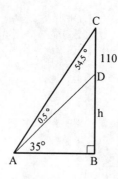

33. We find the missing angle first:

$$\angle C = 180° - (53° + 31°)$$

$$= 180° - 84°$$

$$= 96°$$

Then we find the missing sides using the Law of Sines:

$$\frac{18}{\sin 96°} = \frac{a}{\sin 31°}$$

$$a = \frac{18 \sin 31°}{\sin 96°}$$

$$= \frac{18(0.5150)}{0.9945}$$

$$= 9.3 \text{ miles}$$

$$\frac{18}{\sin 96°} = \frac{b}{\sin 53°}$$

$$a = \frac{18 \sin 53°}{\sin 96°}$$

$$= \frac{18(0.7986)}{0.9945}$$

$$= 14 \text{ miles}$$

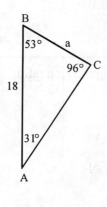

37. We can redraw the two tension vectors $\vec{T}_1$, and $\vec{T}_2$ and the vector $\vec{W}$ due to gravity.

We know that the magnitude of $\vec{W}$ is 1,850 pounds.

First, we find the missing angle:

$$\theta = 180° - (71.8° + 74.8°)$$
$$= 180° - 146.6°$$
$$= 33.4°$$

Then, we find $\vec{T}_1$, and $\vec{T}_2$ using the Law of Sines:

$$\frac{\vec{T}_1}{\sin 71.8°} = \frac{1,850}{\sin 33.4°}$$

$$\vec{T}_1 = \frac{1,850 \sin 71.8°}{\sin 33.4°}$$

$$= \frac{1,850(0.9500)}{0.5505}$$

$$= 3,193 \text{ pounds}$$

$$\frac{\vec{T}_2}{\sin 74.8°} = \frac{1,850}{\sin 33.4°}$$

$$\vec{T}_2 = \frac{1,850 \sin 74.8°}{\sin 33.4°}$$

$$= \frac{1,850(0.9650)}{0.5505}$$

$$= 3,243 \text{ pounds}$$

41.
$$\sin\theta \cos\theta - 2\cos\theta = 0$$

$\cos\theta(\sin\theta - 2) = 0$	Factor
$\cos\theta = 0 \quad \text{or} \quad \sin\theta = 2$	Set each factor $= 0$
$\theta = 90° \text{ or } 270°$	Solve each equation

$$\text{No Solution}$$

45.

$\cos^2\theta - 4\cos\theta + 2 = 0$	Standard form
$\cos\theta = \dfrac{-(-4) \pm \sqrt{(-4)^2 - 4(1)(2)}}{2(1)}$	Quadratic formula: $a = 1$, $b = -4$, and $c = 2$

Continued on next page.

$$= \frac{4 \pm \sqrt{8}}{2} \qquad \text{Simplify}$$

$$= \frac{4 \pm 2\sqrt{2}}{2}$$

$$= 2 \pm \sqrt{2}$$

Then $\cos \theta = 2 + 1.4142$ or $\cos \theta = 2 - 1.4142$

$\cos \theta = 3.4142 \qquad\qquad \cos \theta = 0.5858$

No solution $\qquad\qquad\qquad \theta = 54.1°$ or $305.9°$

49. $\sin \theta = 0.7380$

     $\theta = 47.6°$ or $132.4° \qquad \widehat{\theta} = 47.6°$ and θ is in QI or QII

Problem Set 7.2

1. $\sin B = \dfrac{b \sin A}{a}$

$$= \frac{40 \sin 30°}{10}$$

$$= \frac{40(0.5)}{10}$$

$$= 2$$

Since $\sin B$ can never be greater than 1, no triangle exists.

5. $\sin B = \dfrac{b \sin A}{a}$

$$= \frac{18 \sin 60°}{16}$$

$$= \frac{18(0.8660)}{16}$$

$$= 0.9743$$

$B = 77°$ or $B' = 180° - 77°$

$$B' = 103°$$

Therefore, there are 2 possible triangles.

9. $\sin B = \dfrac{b \sin A}{a}$

$\qquad = \dfrac{22.3 \sin 112.2°}{43.8}$

$\qquad = \dfrac{22.3(0.9259)}{43.8}$

$\qquad = 0.4714$

$B = 28.1°$ or $B' = 180° - 28.1°$

$\qquad\qquad\qquad\qquad = 151.9°$

B' cannot equal 151.9° because $A = 112.2°$.

There can only be one obtuse angle in a triangle.

$\qquad C = 180° - (A + B)$

$\qquad\quad = 180° - (112.2° + 28.1°)$

$\qquad\quad = 180° - 140.3°$

$\qquad\quad = 39.7°$

$\dfrac{c}{\sin C} = \dfrac{a}{\sin A}$

$\qquad c = \dfrac{a \sin C}{\sin A}$

$\qquad\quad = \dfrac{43.8 \sin 39.7°}{\sin 112.2°}$

$\qquad\quad = \dfrac{43.8(0.6388)}{0.9259}$

$\qquad\quad = 30.2$ cm

13. $\sin C = \dfrac{c \sin B}{b}$

$\qquad = \dfrac{1.12 \sin 45° \, 10'}{1.79}$

$\qquad = \dfrac{1.12(0.7092)}{1.79}$

$\qquad = 0.4437$

$C = 26.3°$ or $C' = 180° - 26.3°$

$C = 26° \, 20'$ $C' = 153.7°$

$\qquad\qquad\qquad C' \neq 153.7°$ because $B + C' = 45.2° + 153.7° = 198.9°$ which is impossible.

Continued on next page.

$$A = 180° - (B + C)$$
$$= 180° - (45° \, 10' + 26° \, 20')$$
$$= 180° - 71° \, 30'$$
$$= 108° \, 30'$$

$$a = \frac{b \sin A}{\sin B}$$

$$= \frac{1.79 \sin 108° \, 30'}{\sin 45° \, 10'}$$

$$= \frac{1.79(0.9483)}{0.7092}$$

$$= 2.39 \text{ inches}$$

17. $\sin B = \dfrac{b \sin A}{a}$

$$= \frac{2.9 \sin 142°}{1.4}$$

$$= \frac{2.9(0.6157)}{1.4}$$

$$= \frac{2.9(0.6157)}{1.4}$$

$$= 1.2753$$

Since $\sin B$ can never be greater than 1, no triangle exists.

21. $\sin C = \dfrac{c \sin A}{a}$

$$= \frac{50 \sin 58°}{44}$$

$$= \frac{50(0.8480)}{44}$$

$$= 0.9637$$

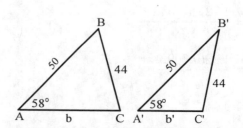

$$C = 75° \quad \text{or} \quad C' = 180° - 75° = 105°$$

$B = 180° - (A + C)$ $B' = 180° - (A' + C')$

$\quad = 180° - (58° + 75°)$ $= 180° - (58° + 105°)$

$\quad = 180° - 33°$ $= 180° - 163°$

$\quad = 47°$ $= 17°$

$$b = \frac{c \sin B}{\sin C} \qquad\qquad b' = \frac{c' \sin B'}{\sin C'}$$

$$= \frac{44 \sin 47°}{\sin 58°} \qquad\qquad = \frac{44 \sin 17°}{\sin 58°}$$

$$= \frac{44(0.7314)}{0.8480} \qquad\qquad = \frac{44(0.2924)}{0.8480}$$

$$= 38 \text{ feet} \qquad\qquad = 15 \text{ feet}$$

25. $\sin B = \dfrac{b \sin A}{a}$

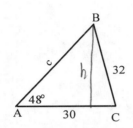

$$= \frac{30 \sin 48°}{32}$$

$$= \frac{30(0.7431)}{32}$$

$$= 0.5772$$

$$B = 44° \quad \text{or} \quad B' = 180° - 44°$$

$$= 136° \text{ This is impossible since the sum of the angles}$$

would be greater than 180°.

$$C = 180° - (48° + 44°) = 88°$$

29.
$$2 \cos \theta - \sin 2\theta = 0$$
$$2 \cos \theta - 2 \sin \theta \cos \theta = 0$$
$$2 \cos \theta (1 - \sin \theta) = 0$$

$$2 \cos \theta = 0 \qquad \text{or} \qquad 1 - \sin \theta = 0$$
$$\cos \theta = 0 \qquad\qquad\qquad \sin \theta = 1$$
$$\theta = 90° \text{ or } 270° \qquad\qquad \theta = 90°$$

33.
$$2 \cos x - \sec x + \tan x = 0$$
$$2 \cos x - \frac{1}{\cos x} + \frac{\sin x}{\cos x} = 0$$
$$2 \cos^2 x - 1 + \sin x = 0$$
$$2(1 - \sin^2 x) - 1 + \sin x = 0$$
$$2 - 2 \sin^2 x - 1 + \sin x = 0$$
$$2 \sin^2 x - \sin x - 1 = 0$$

Continued on next page.

$$(2 \sin x + 1)(\sin x - 1) = 0$$

$$2 \sin x + 1 = 0 \quad \text{or} \quad \sin x - 1 = 0$$

$$2 \sin x = -1 \qquad\qquad \sin x = 1$$

$$\sin x = -\frac{1}{2} \qquad\qquad x = \frac{\pi}{2} + 2k\pi$$

$$x = \frac{7\pi}{6} + 2k\pi \text{ or } \frac{11\pi}{6} + 2k\pi$$

Problem Set 7.3

1. $c^2 = a^2 + b^2 - 2ab \cos C$

    $= (120)^2 + (66)^2 - 2(120)(66) \cos 60°$

    $= 14,400 + 4,356 - 15,840(0.5)$

    $= 14,400 + 4,356 - 7,920$

    $= 10,836$

    $c = 100$ inches (rounded to 2 significant digits)

5. $a^2 = b^2 + c^2 - 2bc \cos A$

    $= (4.2)^2 + (6.8)^2 - 2(4.2)(6.8) \cos 116°$

    $= 17.64 + 46.24 - 57.12(-0.4384)$

    $= 17.64 + 46.24 + 25.04$

    $= 88.92$

    $a = 9.4$ meters

9. $b^2 = a^2 + c^2 - 2ac \cos B$

    $= (410)^2 + (340)^2 - 2(410)(340) \cos 151.5°$

    $= 168,100 + 115,600 - 278,800(-0.8788)$

    $= 528,714.211$

    $b = 727$ meters

    $\sin A = \dfrac{a \sin B}{b}$

    $= \dfrac{410 \sin 151.5°}{727}$

    $= \dfrac{410(0.4772)}{727}$

    $= 0.2691$

    $A = 15.6°$

$$C = 180° - (A + B)$$
$$= 180° - (15.6° + 151.5°)$$
$$= 180° - (167.1°)$$
$$= 12.9°$$

13. $a^2 = b^2 + c^2 - 2bc \cos A$

$$= (0.923)^2 + (0.387)^2 - 2(0.923)(0.387) \cos 43° 20'$$
$$= 0.851929 + 0.149769 - 0.714402(0.7274)$$
$$= 0.4821$$

$a = 0.694$ kilometers

$$\sin C = \frac{c \sin A}{a}$$
$$= \frac{0.387 \sin 43° 20'}{0.694}$$
$$= \frac{0.387(0.6862)}{0.694}$$
$$= 0.3827$$

$$C = 22° 30'$$
$$B = 180° - (A + C)$$
$$= 180° - (22° 30' + 43° 20')$$
$$= 180° - (65° 50')$$
$$= 114° 10'$$

17. $a^2 = b^2 + c^2 - 2bc \cos A$

$a^2 = b^2 + c^2 - 2bc(\cos 90°)$

$a^2 = b^2 + c^2 - 2bc(0)$

$a^2 = b^2 + c^2$

21. $d_1 = r_1 t_1$ $d_2 = r_2 t_2$

$= 130(1.5)$ $= 150 (.15)$

$= 195$ miles $= 225$ miles

$a^2 = b^2 + c^2 - 2bc \cos A$

$= (225)^2 + (195)^2 - 2(225)(195) \cos 36°$

$= 17658.76$

$a = 130$ miles (rounded to 2 significant digits)

25.
$$|\vec{V} + \vec{W}|^2 = |\vec{V}|^2 + |\vec{W}|^2 - 2|\vec{V}||\vec{W}|\cos\theta$$
$$= (35)^2 + (160)^2 - 2(35)(160)\cos 165°$$
$$= 1{,}225 + 25{,}600 - 11{,}200(-0.9659)$$
$$= 1{,}225 + 25{,}600 + 10{,}818$$
$$= 37{,}643$$

$$|\vec{V} + \vec{W}| = 190 \text{ mph (to two significant digits)}$$

$$\frac{\sin\beta}{35} = \frac{\sin 165°}{190}$$

$$\sin\beta = \frac{35\sin 165°}{190}$$

$$= \frac{35(0.2588)}{190}$$

$$= 0.0477$$

$$\beta = 3° \text{ (to the nearest degree)}$$

The true course is $150° + \beta = 150° + 3° = 153°$. The speed of the plane with respect to the ground is 190 miles per hour.

29. $\sin 3x = \dfrac{1}{2}$

$$3x = \frac{\pi}{6} + 2k\pi \quad \text{or} \quad 3x = \frac{5\pi}{6} + 2k\pi$$

$$x = \frac{\pi}{18} + \frac{2k\pi}{3} \qquad x = \frac{5\pi}{18} + \frac{2k\pi}{3}$$

33.
$$2\cos^2 3\theta - 9\cos 3\theta + 4 = 0$$
$$(2\cos 3\theta - 1)(\cos 3\theta - 4) = 0$$
$$2\cos 3\theta - 1 = 0 \quad \text{or} \quad \cos 3\theta - 4 = 0$$
$$2\cos 3\theta = 1 \qquad\qquad \cos 3\theta = 4$$
$$\text{No solution}$$

$$\cos 3\theta = \frac{1}{2}$$

$$3\theta = 60° + 360°k \quad \text{or} \quad 3\theta = 300° + 360°k$$
$$\theta = 20° + 120°k \quad \text{or} \quad \theta = 100° + 120°k$$

37.
$$\sin\theta + \cos\theta = 1$$
$$(\sin\theta + \cos\theta)^2 = 1^2$$
$$\sin^2\theta + 2\sin\theta\cos\theta + \cos^2\theta = 1$$
$$(\sin^2\theta + \cos^2\theta) + 2\sin\theta\cos\theta = 1$$

$$1 + \sin 2\theta = 1$$
$$\sin 2\theta = 0$$
$$2\theta = 0° \text{ or } 180°$$
$$\theta = 0° \text{ or } 90°$$

Check: $\sin 0° + \cos 0° \overset{?}{=} 1$ $\qquad$ $\sin 90° + \cos 90° \overset{?}{=} 1$

$$0 + 1 = 1 \qquad\qquad 1 + 0 = 1$$
$$1 = 1 \qquad\qquad\qquad 1 = 1$$

Problem Set 7.4

1. $S = \dfrac{1}{2}\, ab \sin C$ $\qquad$ Formula for area of a triangle

    $= \dfrac{1}{2}(50)(70) \sin 60°$ $\qquad$ Substitute known values

    $= 1,750\,(0.8660)$ $\qquad$ Simplify

    $= 1,520 \text{ cm}^2$ $\qquad$ Round to 3 significant digits

5. $S = \dfrac{1}{2}\, bc \sin A$ $\qquad$ Formula for area of a triangle

    $= \dfrac{1}{2}(0.923)(0.387) \sin 43°\ 20'$ $\qquad$ Substitute known values

    $= (0.1786)(0.6862)$ $\qquad$ Simplify

    $= 0.123 \text{ km}^2$ $\qquad$ Round to 3 significant digits

6. $C = 180° - (A + B)$

    $= 180° - (42.5° + 71.4°)$

    $= 180° - 113.9°$

    $= 66.1°$

    $S = \dfrac{a^2 \sin B \sin C}{2 \sin A}$ $\qquad$ Formula for area of a triangle

    $= \dfrac{(210)^2 (\sin 71.4°)(\sin 66.1°)}{2 \sin 42.5°}$ $\qquad$ Substitute known values

    $= \dfrac{(44,100)(0.9478)(0.9143)}{2(0.6756)}$ $\qquad$ Simplify

    $= 28,300 \text{ in}^2$ $\qquad$ Round to 3 significant digits

13.

$$s = \frac{1}{2}(a + b + c)$$ Formula for half the perimeter

$$= \frac{1}{2}(44 + 66 + 88)$$ Substitute known values

$$= 99$$ Simplify

$$S = \sqrt{s(s - a)(s - b)(s - c)}$$ Formula for area of a triangle

$$= \sqrt{99(99 - 44)(99 - 66)(99 - 88)}$$ Substitute known values

$$= \sqrt{99(55)(33)(11)}$$ Simplify

$$= \sqrt{1,976,535}$$

$$= 1,410 \text{ in}^2$$ Round to 3 significant digits

17.

$$s = \frac{1}{2}(a + b + c)$$ (Explanation same as Problem 13)

$$= \frac{1}{2}(4.38 + 3.79 + 5.22)$$

$$= 6.695$$

$$S = \sqrt{s(s - a)(s - b)(s - c)}$$

$$= \sqrt{6.695(6.695 - 4.38)(6.695 - 3.79)(6.695 - 5.22)}$$

$$= \sqrt{6.695(2.315)(2.905)(1.475)}$$

$$= \sqrt{66.41}$$

$$= 8.15 \text{ ft}^2$$

21.

$$C = 180° - (A + B)$$

$$= 180° - (30° + 50°)$$

$$= 180° - 80°$$

$$= 100°$$

$$S = \frac{c^2 \sin A \sin B}{2 \sin C}$$ Formula for area of a triangle

$$c^2 = \frac{2S \sin C}{\sin A \sin B}$$ Multiply both sides by $\dfrac{2 \sin C}{\sin A \sin B}$

$$= \frac{2(40) \sin 100°}{\sin 30° \sin 50°}$$ Substitute known values

$$= \frac{80(0.9848)}{0.5(0.7660)} \qquad \text{Simplify}$$

$$= 205.69$$

$$c = 14.3 \text{ cm} \qquad \text{Round to 3 significant digits}$$

25.
$$3 + 2\sin t = x \qquad\qquad 1 + 2\cos t = y$$

$$2\sin t = x - 3 \qquad\qquad 2\cos t = y - 1$$

$$\sin t = \frac{x - 3}{2} \qquad\qquad \cos t = \frac{y - 1}{2}$$

$$\sin^2 t + \cos^2 t = 1$$

$$\left(\frac{x - 3}{2}\right)^2 + \left(\frac{y - 1}{2}\right)^2 = 1$$

$$\frac{(x - 3)^2}{4} + \frac{(y - 1)^2}{4} = 1$$

$$(x - 3)^2 + (y - 1)^2 = 4$$

The graph is a circle with center at (3, 1) and radius of 2.

29.
$$\cos 2t = y \quad \text{and} \quad \sin t = x$$

$$\cos 2t = 1 - 2\sin^2 t$$

$$y = 1 - 2x^2$$

Problem Set 7.5

13.
$$|\vec{U}| = \sqrt{a^2 + b^2}$$

$$= \sqrt{5^2 + (-12)^2}$$

$$= \sqrt{25 + 144}$$

$$= \sqrt{169}$$

$$= 13$$

17. Let $\vec{V} = <-5, 6>$

$$|\vec{V}| = \sqrt{(-5)^2 + 6^2}$$

$$= \sqrt{25 + 36}$$

$$= \sqrt{61}$$

21. Let $\vec{V} = <-2, -5>$

$$\vec{|V|} = \sqrt{(-2)^2 + (-5)^2}$$
$$= \sqrt{4 + 25}$$
$$= \sqrt{29}$$

25.

$$\vec{U} + \vec{V} = 6\vec{i} + (-8\vec{j}) \qquad \vec{U} - \vec{V} = 6\vec{i} - (-8\vec{j})$$
$$= 6\vec{i} - 8\vec{j} \qquad\qquad = 6\vec{i} + 8\vec{j}$$

$$3\vec{U} + 2\vec{V} = 3(6\vec{i}) + 2(-8\vec{j})$$
$$= 18\vec{i} - 16\vec{j}$$

$$\vec{U} \bullet \vec{V} = (6\vec{i} + 0\vec{j}) \bullet (0\vec{i} - 8\vec{j})$$
$$= 6(0) + 0(-8)$$
$$= 0$$

29.

$$<6, 6> \bullet <3, 5> = 6(3) + 6(5)$$
$$= 18 + 30$$
$$= 48$$

33. First, we must find $\vec{U} \bullet \vec{V}$, $\vec{|U|}$, and $\vec{|V|}$:

$$\vec{U} \bullet \vec{V} = <13, 0> \bullet <0, -6>$$
$$= 13(0) + 0(-6)$$
$$= 0$$

$$\vec{|U|} = 13$$
$$\vec{|V|} = 6$$

$$\cos\theta = \frac{\vec{U} \bullet \vec{V}}{\vec{|U|}\,\vec{|V|}}$$
$$= \frac{0}{13 \cdot 6}$$
$$= \frac{0}{78}$$
$$= 0$$
$$\theta = 90.0°$$

37. First, we must find $\vec{U} \bullet \vec{V}$, $|\vec{U}|$, and $|\vec{V}|$:

$$\vec{U} \bullet \vec{V} = <13, -8> \bullet <2, 11>$$
$$= 13(2) + (-8)(11)$$
$$= 26 - 88$$
$$= -62$$

$$|\vec{U}| = \sqrt{13^2 + (-8)^2} \qquad\qquad |\vec{V}| = \sqrt{2^2 + 11^2}$$
$$= \sqrt{169 + 64} \qquad\qquad = \sqrt{4 + 121}$$
$$= \sqrt{233} \qquad\qquad = \sqrt{125}$$

$$\cos\theta = \frac{\vec{U} \bullet \vec{V}}{|\vec{U}|\,|\vec{V}|}$$
$$= \frac{-62}{\sqrt{233}\,\sqrt{125}}$$
$$= -0.3633$$
$$\theta = 111.3°$$

41. To show that $-\vec{i}$ and $\vec{j}$ are perpendicular, we must show that their dot product equals zero.

$$-\vec{i} = <-1, 0> \text{ and } \vec{j} = <0, 1>$$
$$-\vec{i} \bullet \vec{j} = -1(0) + 0(1)$$
$$= 0 + 0$$
$$= 0$$

Therefore, they are perpendicular.

45. Let $\vec{U} = a\vec{i} + b\vec{j}$ and $\vec{V} = c\vec{i} + d\vec{j}$

Then $\vec{U} - \vec{V} = (a - c)\vec{i} + (b - d)\vec{j}$

$$|\vec{U} - \vec{V}|^2 = (a - c)^2 + (b - d)^2$$
$$= a^2 - 2ac + c^2 + b^2 - 2bd + d^2$$
(1) $$= a^2 + b^2 + c^2 + d^2 - 2(ac + bd)$$

Using the Law of Cosines, we also have:

(2) $$|\vec{U} - \vec{V}|^2 = |\vec{U}|^2 + |\vec{V}|^2 - 2|\vec{U}|\,|\vec{V}| \cos\theta$$

Continued on next page.

We know that $|\vec{U}|^2 = a^2 + b^2$, $|\vec{V}|^2 = c^2 + d^2$, and $\vec{U} \bullet \vec{V} = ac + bd$

Substituting these values into (2), we get:

(3) $|\vec{U} - \vec{V}|^2 = a^2 + b^2 + c^2 + d^2 - 2\,|\vec{U}|\,|\vec{V}|\cos\theta$

Setting (1) equal to (3) and simplifying, we get:

$$a^2 + b^2 + c^2 + d^2 - 2(ac + bd) = a^2 + b^2 + c^2 + d^2 - 2|\vec{U}|\,\vec{V}\,\cos\theta$$

$$-2(ac + bd) = -2\,|\vec{U}|\,|\vec{V}|\cos\theta$$

$$ac + bd = |\vec{U}|\,|\vec{V}|\cos\theta$$

$$\vec{U} \bullet \vec{V} = |\vec{U}|\,|\vec{V}|\cos\theta$$

49.   We redraw the vectors and find the missing angles:

$$\theta = 90° - 25.5°$$

$$= 64.5°$$

Then $\vec{H} = -|\vec{H}|\,\vec{i}$

$\vec{W} = -95.5\,\vec{j}$

$\vec{T} = |\vec{T}|\cos 64.5°\vec{i} + |\vec{T}|\sin 64.5°\vec{j}$

and $\vec{H} + \vec{W} + \vec{T} = 0$

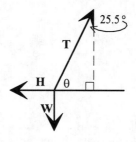

Collecting all the $\vec{i}$ components and all the $\vec{j}$ components together, we have:

$(-|\vec{H}| + |\vec{T}|\cos 64.5°)\,\vec{i} + (-95.5 + |\vec{T}|\sin 64.5°)\,\vec{j} = 0$

Therefore, $-|\vec{H}| + |\vec{T}|\cos 64.5° = 0$ and $-95.5 + |\vec{T}|\sin 64.5° = 0$

Solving the second equation for $|\vec{T}|$, we get:

$$|\vec{T}|\sin 64.5° = 95.5$$

$$|\vec{T}| = \frac{95.5}{\sin 64.5°}$$

$$= \frac{95.5}{0.9026}$$

$$= 105.807$$

We substitute this into the first equation and solve for $|\vec{H}|$:

$$-|\vec{H}| + 105.8 \cos 64.5° = 0$$

$$|\vec{H}| = 105.807 \cos 64.5°$$

$$= 105.807(0.4305)$$

$$= 46.5 \text{ lbs.}$$

Chapter 7 Test

1. $b = \dfrac{a \sin B}{\sin A}$

   $= \dfrac{3.8 \sin 70°}{\sin 32°}$

   $= \dfrac{3.8(0.9397)}{0.5299}$

   $= 6.7$ in

2. $b = \dfrac{c \sin B}{\sin C}$

   $= \dfrac{2.9 \sin 118°}{\sin 37°}$

   $= \dfrac{2.9(0.8829)}{0.6018}$

   $= 4.3$ in

3. $C = 180° - (A + B)$

   $= 180° - (38.2° + 63.4°)$

   $= 180° - 101.6°$

   $= 78.4°$

   $a = \dfrac{c \sin A}{\sin C}$ $b = \dfrac{c \sin B}{\sin c}$

   $= \dfrac{42.0 \sin 38.2°}{\sin 78.4°}$ $= \dfrac{42.0 \sin 63.4°}{\sin 78.4°}$

   $= \dfrac{42.0(0.6184)}{0.9796}$ $= \dfrac{42.0(0.8942)}{0.9796}$

   $= 26.5$ cm $= 38.3$ cm

4. $B = 180° - (A + C)$

   $= 180° - (24.7° + 106.1°)$

   $= 180° - 130.8°$

   $= 49.2°$

Continued on next page.

$$a = \frac{b \sin A}{\sin B} \qquad\qquad c = \frac{b \sin C}{\sin B}$$

$$= \frac{34.0 \sin 24.7°}{\sin 49.2°} \qquad\qquad = \frac{34.0 \sin 106.1°}{\sin 49.2°}$$

$$= \frac{34.0(0.4179)}{0.7570} \qquad\qquad = \frac{34.0(0.9608)}{0.7570}$$

$$= 18.8 \text{ cm} \qquad\qquad = 43.2 \text{ cm}$$

5.　$$\sin B = \frac{b \sin A}{a}$$

$$= \frac{42 \sin 60°}{12}$$

$$= \frac{42(0.8660)}{12}$$

$$= 3.0311$$

Since $\sin B$ is never greater than 1, no triangle exists.

6.　$$\sin B = \frac{b \sin A}{a}$$

$$= \frac{21 \sin 42°}{29}$$

$$= \frac{21(0.6691)}{29}$$

$$= 0.4845$$

$$B = 29° \quad \text{or} \quad B' = 180° - 29°$$

$$= 151° \quad \text{This is impossible: the sum of the angles is greater than } 180°.$$

Therefore, the only possibility is $B = 29°$ which means that only one triangle exists.

7.　$$\sin B = \frac{b \sin A}{a}$$

$$= \frac{7.9 \sin 51°}{6.5}$$

$$= \frac{7.9(0.7771)}{6.5}$$

$$= 0.9445$$

$$B = 71° \quad \text{or} \quad B' = 180° - 71°$$

$$= 109°$$

$$C = 180° - (51° + 71°)$$
$$= 180° - 122°$$
$$= 58°$$

$$C' = 180° - (51° + 109°)$$
$$= 180° - 160°$$
$$= 20°$$

$$c = \frac{a \sin C}{\sin A}$$
$$= \frac{6.5 \sin 58°}{\sin 51°}$$
$$= \frac{6.5(0.8480)}{0.7771}$$
$$= 7.1 \text{ ft}$$

$$c' = \frac{a \sin C'}{\sin A}$$
$$= \frac{6.5 \sin 20°}{\sin 51°}$$
$$= \frac{6.5(0.3420)}{0.7771}$$
$$= 2.9 \text{ ft}$$

8.
$$\sin B = \frac{b \sin A}{a}$$
$$= \frac{9.4 \sin 26°}{4.8}$$
$$= \frac{9.4(0.4384)}{4.8}$$
$$= 0.8585$$
$$B = 59° \qquad \text{or}$$

$$B' = 180° - 59°$$
$$= 121°$$

$$C = 180° - (26° + 59°)$$
$$= 180° - 85°$$
$$= 95°$$

$$C' = 180° - (26° + 121°)$$
$$= 180° - 147°$$
$$= 33°$$

$$c = \frac{a \sin C}{\sin A}$$
$$= \frac{4.8 \sin 95°}{\sin 26°}$$
$$= \frac{4.8(0.9962)}{0.4384}$$
$$= 11 \text{ ft}$$

$$c' = \frac{a \sin C'}{\sin A}$$
$$= \frac{4.8 \sin 33°}{\sin 26°}$$
$$= \frac{4.8(0.5446)}{0.4384}$$
$$= 6.0 \text{ ft}$$

9.
$$c^2 = a^2 + b^2 - 2ab\cos C$$
$$= (10)^2 + (12)^2 - 2(10)(12)\cos 60°$$
$$= 100 + 144 - 240(0.5)$$
$$= 124$$
$$c = 11 \text{ cm}$$

10.
$$c^2 = a^2 + b^2 - 2ab\cos C$$
$$= 10^2 + 12^2 - 2(10)(12)\cos 120°$$
$$= 100 + 144 - 240(-0.5)$$
$$= 364$$
$$c = 19 \text{ cm}$$

11.
$$\cos C = \frac{a^2 + b^2 - c^2}{2ab}$$
$$= \frac{5^2 + 7^2 - 9^2}{2(5)(7)}$$
$$= \frac{-7}{70}$$
$$= -0.1$$
$$C = 95.7°$$

12.
$$\cos B = \frac{a^2 + c^2 - b^2}{2ac}$$
$$= \frac{10^2 + 11^2 - 12^2}{2(10)(11)}$$
$$= \frac{77}{220}$$
$$= 0.35$$
$$B = 69.5°$$

13.
$$c^2 = a^2 + b^2 - 2ab\cos C$$
$$= (6.4)^2 + (2.8)^2 - 2(6.4)(2.8)\cos 119°$$
$$= 40.96 + 7.84 - 35.84(-0.4848)$$
$$= 66.18$$
$$c = 8.1 \text{ cm}$$

$$\sin B = \frac{b \sin C}{c}$$

$$= \frac{2.8 \sin 119°}{8.1}$$

$$= \frac{2.8(0.8746)}{8.1}$$

$$= 0.3023$$

$$B = 18°$$

$$A = 180° - (B + C)$$

$$= 180° - (18° + 119°)$$

$$= 180° - 137°$$

$$= 43°$$

14. $a^2 = b^2 + c^2 - 2bc \cos A$

$$= (3.7)^2 + (6.2)^2 - 2(3.7)(6.2)\cos 35°$$

$$= 13.69 + 38.44 - 45.88(0.8192)$$

$$= 14.547$$

$$a = 3.8 \text{ m}$$

$$\sin B = \frac{b \sin A}{a}$$

$$= \frac{3.7 \sin 35°}{3.8}$$

$$= \frac{3.7(0.5736)}{3.8}$$

$$= 0.5585$$

$$B = 34°$$

$$C = 180° - (35° + 34°)$$

$$= 180° - 69°$$

$$= 111°$$

15.
$$\cos\theta = \frac{38^2 + 48^2 - 38^2}{2(38)(48)}$$

$$= \frac{2{,}304}{3{,}648}$$

$$= 0.6316$$

$$\theta = 51°$$

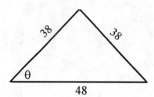

16.
$$\tan 48° = \frac{h}{53}$$

$$h = 53\tan 48°$$

$$= 53(1.1106)$$

$$= 59 \text{ ft}$$

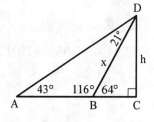

17. First, we find the missing angles of $\triangle$ ABD:

$$\angle ABD = 180° - 64° = 116°$$

$$\angle ADB = 180° - (43° + 116°) = 21°$$

Next, we find x using the Law of Sines:

$$\frac{x}{\sin 43°} = \frac{240}{\sin 21°}$$

$$x = \frac{240\sin 43°}{\sin 21°}$$

$$= \frac{240(0.6820)}{0.3584}$$

$$= 457$$

Then, we find h using the sine relationship:

$$\sin 64° = \frac{h}{x}$$

$$h = x\sin 64°$$

$$= 457(0.8988)$$

$$= 410 \text{ ft (rounded to 2 significant digits)}$$

18. From Geometry, we know that the diagonals of a parallelogram bisect each other.

Since we want to find the shorter side, we must find θ:

$$\theta = 180° - 134.5°$$

$$= 45.5°$$

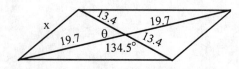

Now we can find x using the Law of Cosines:

$$x^2 = (13.4)^2 + (19.7)^2 - 2(13.4)(19.7)\cos 45.5°$$
$$= 179.56 + 388.09 - 527.96(0.7009)$$
$$= 197.598$$
$$= 14.1 \text{ m}$$

19.
$$CD = r + 4.55$$
$$= 3960 + 4.55$$
$$= 3964.55$$

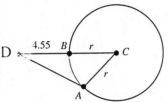

$$\frac{\sin D}{3,960} = \frac{\sin 90.8°}{3,964.55}$$

$$\sin D = \frac{3,960(0.9999)}{3,964.55}$$

$$= 0.9988$$
$$D = 87.1°$$
$$C = 180° - (90.8° + 87.1°)$$
$$= 180° - 177.9°$$
$$= 2.1°$$

To use the arc length formula, we must change 2.1° to radian measure:

$$2.1° = 2.1\left(\frac{\pi}{180}\right)$$
$$= 0.03665$$
$$S = r\theta$$
$$= 3960(0.03665)$$
$$= 145 \text{ miles}$$

20. First, we find angle B, using alternate interior angles of parallel lines:

$$B = 55° + 44° = 99°$$

Next, we find b using the Law of Cosines:

$$b^2 = (2.2)^2 + (3.3)^2 - 2(2.2)(3.3)\cos 99°$$
$$= 4.84 + 10.89 - 14.52(-0.1564)$$
$$= 18.001$$
$$b = 4.2 \text{ mi}$$

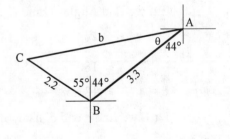

To find his bearing, we must find θ:

$$\sin \theta = \frac{2.2 \sin 99°}{4.2}$$

$$= \frac{2.2(0.9877)}{4.2}$$

$$= 0.5174$$
$$\theta = 31° \qquad \text{The bearing is S 75° W.}$$

21. $C = 180° - (A + B)$

  $= 180° - (47° + 37°)$

  $= 96°$

We find c using the Law of Cosines.

$c^2 = a^2 + b^2 - 2ab \cos C$

  $= (56)^2 + (65)^2 - 2(56)(65) \cos 96°$

  $= 3{,}136 + 4{,}225 - 7{,}280(-0.1045)$

  $= 8{,}122$

$c = 90 \text{ ft}$

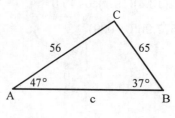

22. First, we find angle A:

  $A = 95.5° - 90°$

  $= 5.5°$

Next, we find angle B using the Law of Sines:

  $\sin B = \dfrac{b \sin A}{a}$

  $= \dfrac{345 \sin 5.5°}{55}$

  $= 0.6012$

  $B = 37°$ or $B' = 180° - 37°$

  $= 143°$

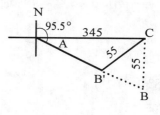

Next we find Angles C and C':

$C = 180° - (5.5° + 37°)$ $C' = 180° - (143° + 5.5°)$

  $= 180° - 42.5°$ $= 180° - 148.5°$

  $= 137.5°$ $= 31.5°$

Last, we find c and c' using the Law of Sines:

$c = \dfrac{a \sin C}{\sin A}$ $c' = \dfrac{a \sin C'}{\sin A}$

  $= \dfrac{55 \sin 137.5°}{\sin 5.5°}$ $= \dfrac{55 \sin 31.5°}{\sin 5.5°}$

  $= \dfrac{55(0.6756)}{0.0958}$ $= \dfrac{55(0.5225)}{0.0958}$

  $= 388 \text{ mph}$ $= 300 \text{ mph}$

23. First, we find the angles of $\triangle ABC$:

$$\alpha = 90° - 48° = 42°$$

$$\beta = 90° - 38° = 52°$$

$$\gamma = 180° - (42° + 52°)$$

$$= 180° - 94°$$

$$= 86°$$

Next we find x using the Law of Sines:

$$x = \frac{25 \sin 52°}{\sin 86°}$$

$$= \frac{25(0.7880)}{0.9976}$$

$$= 19.7$$

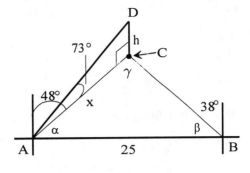

Then we find h:

$$\tan 73° = \frac{h}{19.7}$$

$$h = 19.7 \tan 73°$$

$$= 19.7(3.2709)$$

$$= 65 \text{ ft}$$

Therefore, the tree is 65 ft tall.

24. From Geometry:

$$\alpha = 87.6°$$

$$\beta = 262.6° - 180° = 82.6°$$

$$\alpha + \beta = 170.2°$$

We find x using the Law of Cosines:

$$x^2 = 65.4^2 + 325^2 - 2(65.4)(325)\cos 170.2°$$

$$= 4,277.16 + 105,625 - 42,510(-0.9854)$$

$$= 151,791.85$$

$$x = 390 \text{ mph}$$

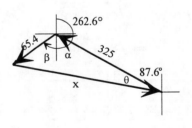

Next we find θ using the Law of Sines:

$$\sin \theta = \frac{65.4 \sin 170.2°}{390}$$

$$= \frac{65.4(0.1702)}{390}$$

$$= 0.0285$$

$$\theta = 1.6°$$

The bearing is N 89.2° W and the ground speed is 390 mph.

---

25. $|\vec{U}| = \sqrt{a^2 + b^2}$

$\qquad = \sqrt{5^2 + 12^2}$

$\qquad = \sqrt{25 + 144}$

$\qquad = \sqrt{169}$

$\qquad = 13$

26. $3\vec{U} + 5\vec{V} = 3(5\vec{i} + 12\vec{j}) + 5(-4\vec{i} + \vec{j})$

$\qquad = 15\vec{i} + 36\vec{j} - 20\vec{i} + 5\vec{j}$

$\qquad = -5\vec{i} + 41\vec{j}$

27. $3\vec{U} - 5\vec{V} = 3(5\vec{i} + 12\vec{j}) - 5(-4\vec{i} + \vec{j})$

$\qquad = 15\vec{i} + 36\vec{j} + 20\vec{i} - 5\vec{j}$

$\qquad = 35\vec{i} + 31\vec{j}$

28. $2\vec{V} - \vec{W} = 2(-4\vec{i} + \vec{j}) - (\vec{i} - 4\vec{j})$

$\qquad = -8\vec{i} + 2\vec{j} - \vec{i} + 4\vec{j}$

$\qquad = -9\vec{i} + 6\vec{j}$

$\qquad |2\vec{V} - \vec{W}| = \sqrt{(-9)^2 + 6^2}$

$\qquad\qquad = \sqrt{81 + 36}$

$\qquad\qquad = \sqrt{117}$

29. $\vec{V} \bullet \vec{W} = (-4\vec{i} + \vec{j}) \bullet (\vec{i} - 4\vec{j})$

$\qquad = -4(1) + 1(-4)$

$\qquad = -4 - 4$

$\qquad = -8$

30. $|\vec{U}| = 13$ from Problem 25

$\qquad |\vec{V}| = \sqrt{(-4)^2 + 1^2}$

$\qquad\qquad = \sqrt{16 + 1}$

$\qquad\qquad = \sqrt{17}$

$$\vec{U} \bullet \vec{V} = (5\vec{i} + 12\vec{j}) \bullet (-4\vec{i} + \vec{j})$$

$$= 5(-4) + 12(1)$$

$$= -20 + 12$$

$$= -8$$

$$\cos\theta = \frac{\vec{U} \bullet \vec{V}}{|\vec{U}|\,|\vec{V}|}$$

$$= \frac{-8}{13\sqrt{17}}$$

$$= -0.1493$$

$$\theta = 98.6°$$

31.

$$S = \frac{c^2 \sin A \sin B}{2 \sin C}$$

$$= \frac{(42.0)^2 \sin 38.2° \sin 63.4°}{2 \sin 78.4°}$$

$$= \frac{1{,}764(0.6184)(0.8942)}{2(0.9796)}$$

$$= 498 \text{ cm}^2$$

32. $S = \frac{1}{2}ab\sin C$ \qquad (See Problem 4)

$$= \frac{1}{2}(18.8)(34.0)\sin 106.1°$$

$$= 319.6(0.9608)$$

$$= 307 \text{ cm}^2$$

33. $S = \frac{1}{2}ab\sin C$

$$= \frac{1}{2}(10)(12)\sin 60°$$

$$= 60(0.8660)$$

$$= 52 \text{ cm}^2$$

34. $S = \frac{1}{2}ab\sin C$

$$= \frac{1}{2}(10)(12)\sin 120°$$

$$= 60(0.8660)$$

$$= 52 \text{ cm}^2$$

35.
$$s = \tfrac{1}{2}(a + b + c)$$
$$= \tfrac{1}{2}(5 + 7 + 9)$$
$$= 10.5$$

$$S = \sqrt{s(s - a)(s - b)(s - c)}$$
$$= \sqrt{10.5(10.5 - 5)(10.5 - 7)(10.5 - 9)}$$
$$= \sqrt{10.5(5.5)(3.5)(1.5)}$$
$$= \sqrt{303.1875}$$
$$= 17 \text{ km}^2$$

36.
$$s = \tfrac{1}{2}(10 + 12 + 11)$$
$$= \tfrac{1}{2}(33)$$
$$= 16.5$$
$$S = \sqrt{16.5(16.5 - 10)(16.5 - 12)(16.5 - 11)}$$
$$= \sqrt{16.5(6.5)(4.5)(5.5)}$$
$$= \sqrt{2,654.4375}$$
$$= 52 \text{ km}^2$$

CHAPTER 8 Complex Numbers and Polar Coordinates

Problem Set 8.1

1. $\sqrt{-16} = i\sqrt{16}$

$\phantom{\sqrt{-16}} = 4i$

5. $\sqrt{-18} = i\sqrt{18}$

$\phantom{\sqrt{-18}} = 3i\sqrt{2}$

9. $\sqrt{-4}\,\sqrt{-9} = i\sqrt{4}\,i\sqrt{9}$

$\phantom{\sqrt{-4}\,\sqrt{-9}} = (2i)(3i)$

$\phantom{\sqrt{-4}\,\sqrt{-9}} = 6i^2$

$\phantom{\sqrt{-4}\,\sqrt{-9}} = 6(-1)$

$\phantom{\sqrt{-4}\,\sqrt{-9}} = -6$

13. $4 = 6x$ and $7 = -14y$

$x = \frac{4}{6} \qquad\qquad y = -\frac{7}{14}$

$ = \frac{2}{3} \qquad\qquad\quad = -\frac{1}{2}$

17. $x^2 - 6 = x$ and $9 = y^2$

$\; x^2 - x - 6 = 0 \qquad\quad y^2 = 9$

$\;(x - 3)(x + 2) = 0 \qquad y = \pm 3$

$\; x - 3 = 0 \;\text{ or }\; x + 2 = 0$

$\qquad x = 3 \qquad\qquad x = -2$

21. $\sin^2 x + 1 = 2\sin x$ and $\tan y = 1$

$\;\sin^2 x - 2\sin x + 1 = 0$

$\;(\sin x - 1)(\sin x - 1) = 0$

$\qquad\quad \sin x - 1 = 0 \qquad\qquad y = \frac{\pi}{4}, \frac{5\pi}{4}$

$\qquad\qquad\quad \sin x = 1$

$\qquad\qquad\qquad\; x = \frac{\pi}{2}$

25. $(6 + 7i) - (4 + i) = 6 + 7i - 4 - i$

$ = (6 - 4) + (7 - 1)i$

$ = 2 + 6i$

29. $(3\cos x + 4i\sin y) + (2\cos x - 7i\sin y)$

$\qquad = (3\cos x + 2\cos x) + (4\sin y - 7\sin y)i$

$\qquad = 5\cos x - 3i\sin y$

33. $(7 - 4i) - [(-2 + i) - (3 + 7i)] = (7 - 4i) - [-2 + i - 3 - 7i]$

$ = 7 - 4i + 2 - i + 3 + 7i$

$ = 12 + 2i$

37. $i^{14} = (i^4)^3 \cdot i^2$

$= 1(-1)$

$= -1$

41. $i^{33} = (i^4)^8 \cdot i$

$= 1 \cdot i$

$= i$

45. $(2 - 4i)(3 + i) = 6 + 2i - 12i - 4i^2$

$= 6 - 10i - 4(-1)$

$= 6 - 10i + 4$

$= 10 - 10i$

49. $(5 + 4i)(5 - 4i) = 25 - 16i^2$

$= 25 - 16(-1)$

$= 25 + 16$

$= 41$

53. $2i(3 + i)(2 + 4i) = 2i(6 + 12i + 2i + 4i^2)$

$= 2i(6 + 14i - 4)$

$= 2i(2 + 14i)$

$= 4i + 28i^2$

$= -28 + 4i$

57. $\dfrac{2i}{3+i} \cdot \dfrac{3-i}{3-i} = \dfrac{6i - 2i^2}{9 - i^2}$

$= \dfrac{6i - 2(-1)}{9 - (-1)}$

$= \dfrac{2 + 6i}{10}$

$= \dfrac{2}{10} + \dfrac{6}{10}i$

$= \dfrac{1}{5} + \dfrac{3}{5}i$

61. $\dfrac{5 - 2i}{i} \cdot \dfrac{i}{i} = \dfrac{5i - 2i^2}{i^2}$

$= \dfrac{5i - 2(-1)}{-1}$

$= \dfrac{2 + 5i}{-1}$

$= -2 - 5i$

65. $z_1 z_2 = (2 + 3i)(2 - 3i)$

$= 4 - 9i^2$

$= 4 - 9(-1)$

$= 4 + 9$

$= 13$

69. $2z_1 + 3z_2 = 2(2 + 3i) + 3(2 - 3i)$

$= 4 + 6i + 6 - 9i$

$= 10 - 3i$

73. $(x + 3i)(x - 3i) = x^2 - 9i^2$

$$= x^2 - 9(-1)$$
$$= x^2 + 9$$

77. $x^2 - 2ax + (a^2 + b^2)$

$$= (a + b\,i)^2 - 2a(a + b\,i) + (a^2 + b^2)$$
$$= a^2 + 2abi + b^2i^2 - 2a^2 - 2abi + a^2 + b^2$$
$$= b^2i^2 + b^2$$
$$= b^2(-1) + b^2$$
$$= -b^2 + b^2$$
$$= 0$$

81. If $2x + y = 4$, then $y = 4 - 2x$.

Substituting this into the second equation, we get:

$$x(4 - 2x) = 8$$
$$4x - 2x^2 = 8$$
$$2x^2 - 4x + 8 = 0$$
$$x^2 - 2x + 4 = 0 \quad \text{where } a = 1, b = -2, c = 4$$

$$x = \frac{-(-2) \pm \sqrt{(-2)^2 - 4(1)(4)}}{2(1)}$$

$$= \frac{2 \pm \sqrt{-12}}{2}$$

$$= \frac{2 \pm 2\,i\sqrt{3}}{2}$$

$$= 1 \pm i\sqrt{3}$$

Now, we find the corresponding $y-$values:

if $x = 1 + i\sqrt{3}$, $\qquad y = 4 - 2(1 + i\sqrt{3})$

$$= 4 - 2 - 2\,i\sqrt{3}$$
$$= 2 - 2\,i\sqrt{3}$$

if $x = 1 - i\sqrt{3}$, $\qquad y = 4 - 2(1 - i\sqrt{3})$

$$= 4 - 2 + 2\,i\sqrt{3}$$
$$= 2 + 2\,i\sqrt{3}$$

85.
$$(a + bi) + (c + di) = (a + c) + (b + d)i$$
$$= (c + a) + (d + b)i$$
$$= (c + di) + (a + bi)$$

Therefore, addition of complex numbers is commutative.

89.
$$r = \sqrt{a^2 + b^2}$$
$$\sin\theta = \frac{b}{\sqrt{a^2 + b^2}} \quad \text{and} \quad \cos\theta = \frac{a}{\sqrt{a^2 + b^2}}$$

93.
$$a^2 = b^2 + c^2 - 2bc\cos A$$
$$= (243)^2 + (157)^2 - 2(243)(157)\cos 73.1°$$
$$= 59{,}049 + 24{,}649 - 76{,}302(0.2907)$$
$$= 61{,}517$$
$$a = 248 \text{ cm}$$

$$\sin C = \frac{c\sin A}{a}$$
$$= \frac{157\sin 73.1°}{248}$$
$$= \frac{157(0.9568)}{248}$$
$$C = 37.3°$$

$$B = 180° - (A + C)$$
$$= 180° - (73.1° + 37.3°)$$
$$= 180° - 110.4°$$
$$= 69.6°$$

Problem Set 8.2

1.
$$|3 + 4i| = \sqrt{3^2 + 4^2}$$
$$= \sqrt{9 + 16}$$
$$= \sqrt{25}$$
$$= 5$$

5.
$$|0 - 5i| = \sqrt{0^2 + (-5)^2}$$
$$= \sqrt{25}$$
$$= 5$$

9. $|-4 - 3i| = \sqrt{(-4)^2 + (-3)^2}$

$= \sqrt{16 + 9}$

$= \sqrt{25}$

$= 5$

13. Opposite of $4i = -4i$

Conjugate of $4i = -4i$

17. Opposite of $-5 - 2i = -(-5 - 2i)$

$= 5 + 2i$

Conjugate of $-5 - 2i = -5 + 2i$

21. $4(\cos 120° + i \sin 120°) = 4\left[-\dfrac{1}{2} + i\left(\dfrac{\sqrt{3}}{2} \right) \right]$

$= -2 + 2i\sqrt{3}$

25. $\cos 315° + i \sin 315° = \dfrac{\sqrt{2}}{2} + i\left(-\dfrac{\sqrt{2}}{2} \right)$

$= \dfrac{\sqrt{2}}{2} - \dfrac{\sqrt{2}}{2}i$

29. $100(\cos 143° + i \sin 143°) = 100[-0.7986 + i(0.6018)]$

$= -79.86 + 60.18i$

33. $10(\cos 342° + i \sin 342°) = 10[0.951 + i(-0.309)]$

$= 9.51 - 3.09i$

37. We have $x = 1$ and $y = -1$, therefore

$r = \sqrt{1^2 + (-1)^2} = \sqrt{2}$

We also know than $\tan \theta = \dfrac{-1}{1}$ and θ is in QIV.

Therefore, $\theta = 315°$.

In trigonometric form, $z = r(\cos \theta + i \sin \theta)$

$= \sqrt{2}\,(\cos 315° + i \sin 315°)$

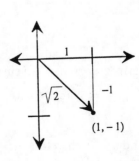

41. We have $x = 0$ and $y = 8$, therefore

$$r = \sqrt{0^2 + 8^2} = 8$$

We also know that $\theta = 90°$.

In trigonometric form, $z = r(\cos\theta + i\sin\theta)$

$$= 8(\cos 90° + i\sin 90°)$$

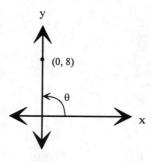

45. We have $x = -2$ and $y = 2\sqrt{3}$, therefore

$r = \sqrt{(-2)^2 + (2\sqrt{3})^2}$
$r = \sqrt{4 + 12}$
$r = \sqrt{16}$
$r = 4$

We also know that $\tan\theta = \frac{2\sqrt{3}}{-2} = -\sqrt{3}$ and θ is in QII.
Therefore, $\theta = 120°$.

In trigonometric form, $z = r(\cos\theta + i\sin\theta)$

$$= 4(\cos 120° + i\sin 120°)$$

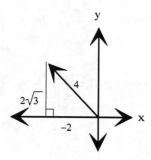

49. We have $x = 20$ and $y = 21$, therefore

$$r = \sqrt{20^2 + 21^2}$$
$$= \sqrt{400 + 441}$$
$$= \sqrt{841}$$
$$= 29$$

We also know that θ is in QI and

$$\tan\theta = \frac{21}{20}$$
$$= 1.05$$
$$\theta = 46.40°$$

In trigonometric form, $z = r(\cos\theta + i\sin\theta)$

$$= 29(\cos 46.40° + i\sin 46.40°)$$

53. We have $x = 11$ and $y = 2$, therefore

$$r = \sqrt{11^2 + 2^2}$$
$$= \sqrt{121 + 4}$$
$$= \sqrt{125}$$
$$= 5\sqrt{5}$$

We also know that θ is in QI and

$$\tan \theta = \frac{2}{11}$$

$$= 0.1818$$

$$\theta = 10.30°$$

In trigonometric form, $z = r(\cos \theta + i \sin \theta)$

$$= 5\sqrt{5}(\cos 10.30° + i \sin 10.30°)$$

57. $\quad 2(\cos 30° + i \sin 30°) = 2\left(\frac{\sqrt{3}}{2} + i \cdot \frac{1}{2}\right) = \sqrt{3} + i$

$2[\cos(-30°) + i \sin(-30°)] = 2[\frac{\sqrt{3}}{2} + i \cdot (-\frac{1}{2})] = \sqrt{3} - i$

61. $\quad \cos 75° = \cos(30° + 45°)$

$$= \cos 30° \cos 45° - \sin 30° \sin 45°$$

$$= \frac{\sqrt{3}}{2} \cdot \frac{\sqrt{2}}{2} - \frac{1}{2} \cdot \frac{\sqrt{2}}{2}$$

$$= \frac{\sqrt{6}}{4} - \frac{\sqrt{2}}{4}$$

$$= \frac{\sqrt{6} - \sqrt{2}}{4}$$

65. $\quad \sin 30° \cos 90° + \cos 30° \sin 90° = \sin(30° + 90°)$

$$= \sin 120°$$

$$= \frac{\sqrt{3}}{2}$$

69. $\quad \sin B = \dfrac{b \sin A}{a}$

$$= \frac{567 \sin 45.6°}{234}$$

$$= \frac{567(0.7145)}{234}$$

$$= 1.7312$$

Since $\sin B$ is never greater than 1, no triangle exists.

Problem Set 8.3

1. $3(\cos 20° + i \sin 20°) \cdot 4(\cos 30° + i \sin 30°)$

$$= 3 \cdot 4[\cos(20° + 30°) + i \sin (20° + 30°)]$$
$$= 12(\cos 50° + i \sin 50°)$$

5. $2(\cos 135° + i \sin 135°) \cdot 2(\cos 45° + i \sin 45°)$

$$= 2 \cdot 2[\cos(135° + 45°) + i \sin(135° + 45°)]$$
$$= 4(\cos 180° + i \sin 180°)$$

9. $z_1 z_2 = (1 + i\sqrt{3})(-\sqrt{3} + i)$

$$= -\sqrt{3} + i - 3i + i^2\sqrt{3}$$
$$= -\sqrt{3} - 2i - \sqrt{3}$$
$$= -2\sqrt{3} - 2i$$

$z_1 = 1 + i\sqrt{3}$ where $x = 1$, $y = \sqrt{3}$ and $r = \sqrt{1^2 + (\sqrt{3})^2} = 2$.

Also, $\tan \theta = \sqrt{3}$ and θ is in QI.

Therefore, $\theta = 60°$.

$z_1 = 2(\cos 60° + i \sin 60°)$ in trigonometric form

$z_2 = -\sqrt{3} + i$ where $x = -\sqrt{3}$, $y = 1$, and $r = \sqrt{(-\sqrt{3})^2 + 1^2} = 2$.

Also, $\tan \theta = -\dfrac{1}{\sqrt{3}}$ and θ is in QII.

Therefore, $\theta = 150°$.

$z_2 = 2(\cos 150° + i \sin 150°)$ in trigonometric form

$z_1 z_2 = 2 \cdot 2[\cos(60° + 150°) + i \sin (60° + 150°)]$

$$= 4(\cos 210° + i \sin 210°)$$

$4(\cos 210° + i \sin 210°) = 4(-\dfrac{\sqrt{3}}{2} - i \cdot \dfrac{1}{2})$

$$= -2\sqrt{3} - 2i$$

13. $z_1 z_2 = (1 + i)(4i)$

$$= 4i + 4i^2$$
$$= 4i + 4(-1)$$
$$= -4 + 4i$$

$z_1 = 1 + i$ where $x = 1$, $y = 1$, and $r = \sqrt{1^2 + 1^2} = \sqrt{2}$.

Also, $\tan \theta = 1$ and θ is in QI. Therefore, $\theta = 45°$.

In trigonometric form, $z_1 = \sqrt{2}(\cos 45° + i \sin 45°)$.

$z_2 = 4i$ where $x = 0$, $y = 4$, and $r = \sqrt{0^2 + 4^2} = 4$

Also, $\theta = 90°$.

In trigonometric form, $z_2 = 4(\cos 90° + i \sin 90°)$.

$$z_1 z_2 = \sqrt{2} \cdot 4[\cos(45° + 90°) + i \sin(45° + 90°)]$$
$$= 4\sqrt{2}(\cos 135° + i \sin 135°)$$

$$4\sqrt{2}(\cos 135° + i \sin 135°) = 4\sqrt{2}(-\tfrac{\sqrt{2}}{2} + i \cdot \tfrac{\sqrt{2}}{2})$$
$$= -4 + 4i$$

17. $$[2(\cos 10° + i \sin 10°)]^6 = 2^6[\cos(6 \cdot 10°) + i \sin(6 \cdot 10°)]$$
$$= 64(\cos 60° + i \sin 60°)$$
$$= 64(\frac{1}{2} + i \cdot \frac{\sqrt{3}}{2})$$
$$= 32 + 32i\sqrt{3}$$

21. $$[3(\cos 60° + i \sin 60°)]^4 = 3^4[\cos(4 \cdot 60°) + i \sin(4 \cdot 60°)]$$
$$= 81(\cos 240° + i \sin 240°)$$
$$= 81[-\frac{1}{2} + i(-\frac{\sqrt{3}}{2})]$$
$$= -\frac{81}{2} - \frac{81\sqrt{3}}{2}i$$

25. First we write $1 + i$ in trigonometric form:

$z = 1 + i$ where $x = 1$, $y = 1$, and $r = \sqrt{1^2 + 1^2} = \sqrt{2}$.

Also, $\tan \theta = 1$ and θ is in QI. Therefore, $\theta = 45°$.

In trigonometric form, $z = \sqrt{2}(\cos 45° + i \sin 45°)$.

$$z^4 = (\sqrt{2})^4[\cos(4 \cdot 45°) + i \sin(4 \cdot 45°)]$$
$$= 4(\cos 180° + i \sin 180°)$$
$$= 4[-1 + i(0)]$$
$$= -4$$

29. First we write $1 - i$ in trigonometric form:

$z = 1 - i$ where $x = 1$, $y = -1$, and $r = \sqrt{1^2 + (-1)^2} = \sqrt{2}$.

Also, $\tan \theta = -1$ and θ is in QIV.

Therefore, $\theta = 315°$.

In trigonometric form, $z = \sqrt{2}(\cos 315° + i \sin 315°)$.

$z^6 = (\sqrt{2})^6[\cos(6 \cdot 315°) + i \sin(6 \cdot 315°)]$

$\quad = 8(\cos 1890° + i \sin 1890°)$

$\quad = 8(\cos 90° + i \sin 90°)$

$\quad = 8(0 + i \cdot 1)$

$\quad = 8i$

33. $\dfrac{20(\cos 75° + i \sin 75°)}{5(\cos 40° + i \sin 40°)} = \dfrac{20}{5}[\cos(75° - 40°) + i \sin(75° - 40°)]$

$\qquad\qquad\qquad\qquad = 4(\cos 35° + i \sin 35°)$

37. $\dfrac{4(\cos 90° + i \sin 90°)}{8(\cos 30° + i \sin 30°)} = \dfrac{4}{8}[\cos(90° - 30°) + i \sin(90° - 30°)]$

$\qquad\qquad\qquad\qquad = \dfrac{1}{2}(\cos 60° + i \sin 60°)$

41. $\dfrac{z_1}{z_2} = \dfrac{\sqrt{3} + i}{2i} \cdot \dfrac{i}{i}$

$\quad = \dfrac{i\sqrt{3} + i^2}{2i^2}$

$\quad = \dfrac{-1 + i\sqrt{3}}{-2}$

$\quad = \dfrac{1}{2} - \dfrac{\sqrt{3}}{2} i$

$z_1 = \sqrt{3} + i$ where $x = \sqrt{3}$, $y = 1$, and $r = \sqrt{(\sqrt{3})^2 + 1^2} = 2$.

Also, $\tan \theta = \dfrac{1}{\sqrt{3}}$ and θ is in QI.

Therefore, $\theta = 30°$.

In trigonometric form, $z_1 = 2(\cos 30° + i \sin 30°)$.

$z_2 = 2i$ where $x = 0$, $y = 2$, and $r = \sqrt{0^2 + 2^2} = 2$.

Also, $\theta = 90°$

In trigonometric form, $z_2 = 2(\cos 90° + i \sin 90°)$.

$$\frac{z_1}{z_2} = \frac{2(\cos 30° + i \sin 30°)}{2(\cos 90° + i \sin 90°)}$$

$$= \frac{2}{2} [\cos(30° - 90°) + i \sin(30° - 90°)]$$

$$= \cos(-60°) + i \sin(-60°)$$

$$\cos(-60°) + i \sin(-60°) = \frac{1}{2} + i(-\frac{\sqrt{3}}{2})$$

$$= \frac{1}{2} - \frac{\sqrt{3}}{2} i$$

45. $\dfrac{z_1}{z_2} = \dfrac{8}{-4} = -2$

$z_1 = 8$ where $x = 8$, $y = 0$, and $r = 8$. Also, $\theta = 0°$

In trigonometric form, $z_1 = 8(\cos 0° + i \sin 0°)$.

$z_2 = -4$ where $x = -4$, $y = 0$, and $r = 4$. Also, $\theta = 180°$.

In trigonometric form, $z_2 = 4(\cos 180° + i \sin 180°)$.

$$\frac{z_1}{z_2} = \frac{8(\cos 0° + i \sin 0°)}{4(\cos 180° + i \sin 180°)}$$

$$= \frac{8}{4} [\cos(0° - 180°) + i \sin(0° - 180°)]$$

$$= 2[\cos(-180°) + \sin(-180°)]$$

$$2[\cos(-180°) + i \sin(-180°)] = 2[-1 + i(0)]$$

$$= -2$$

49. Let $z_1 = 1 + i\sqrt{3}$, where $x = 1$, $y = \sqrt{3}$, and $r = \sqrt{1^2 + (\sqrt{3})^2} = 2$

Also, $\tan \theta = \sqrt{3}$ and θ is in QI.

Therefore, $\theta = 60°$.

In trigonometric form, $z_1 = 2(\cos 60° + i \sin 60°)$.

Let $z_2 = \sqrt{3} - i$, where $x = \sqrt{3}$, $y = -1$, and $r = \sqrt{(\sqrt{3}) + (-1)^2} = 2$

Also, $\tan \theta = -\dfrac{1}{\sqrt{3}}$ and θ is in QIV. Therefore, $\theta = 330°$.

Continued on next page.

In trigonometric form, $z_2 = 2(\cos 330° + i \sin 330°)$.

Let $z_3 = 1 - i\sqrt{3}$, where $x = 1$, $y = -\sqrt{3}$, and $r = \sqrt{1^2 + (-\sqrt{3})^2} = 2$

Also, $\tan\theta = -\sqrt{3}$ and θ is in QIV. Therefore, $\theta = 300°$.

In trigonometric form, $z_3 = 2(\cos 300° + i \sin 300°)$.

Now, we find $\dfrac{(z_1)^4\,(z_2)^2}{(z_3)^3}$:

$$\frac{(z_1)^4\,(z_2)^2}{(z_3)^3} = \frac{2^4[\cos(4 \cdot 60°) + i \sin(4 \cdot 60°)]\,2^2[\cos(2 \cdot 330°) + i \sin(2 \cdot 330°)]}{2^3[\cos(3 \cdot 300°) + i \sin(3 \cdot 300°)]}$$

$$= \frac{16 \cdot 4[\cos(240° + 660°) + i \sin(240° + 660°)]}{8(\cos 900° + i \sin 900°)}$$

$$= \frac{64}{8} \cdot [\cos(900° - 900°) + i \sin(900° - 900°)]$$

$$= 8(\cos 0° + i \sin 0°)$$

$$= 8(1 + i \cdot 0)$$

$$= 8$$

53. $w^4 = [2(\cos 15° + i \sin 15°)]^4$

$$= 2^4[\cos(4 \cdot 15°) + i \sin(4 \cdot 15°)]$$

$$= 16(\cos 60° + i \sin 60°)$$

$$= 16\left(\frac{1}{2} + i \cdot \frac{\sqrt{3}}{2}\right)$$

$$= 8 + 8i\sqrt{3}$$

57. First, we write $\sqrt{3} - i$ in trigonometric form:

$z = \sqrt{3} - i$ where $x = \sqrt{3}$, $y = -1$, and $r = \sqrt{(\sqrt{3})^2 + (-1)^2} = 2$

Also, $\tan\theta = -\dfrac{1}{\sqrt{3}}$ and θ is in QIV.

Therefore, $\theta = 330°$.

In trigonometric form, $z = 2(\cos 330° + i \sin 330°)$.

$z^{-1} = 2^{-1}[\cos(-1)(330°) + i \sin(-1)(330°)]$

$$= \frac{1}{2}[\cos(-330°) + i \sin(-330°)]$$

$$= \frac{1}{2}\left[\frac{\sqrt{3}}{2} + i \cdot \frac{1}{2}\right]$$

$$= \frac{\sqrt{3}}{4} + \frac{1}{4}i$$

61. $\sin \dfrac{A}{2} = \sqrt{\dfrac{1 - \cos A}{2}}$

$$= \sqrt{\dfrac{1 - (-\frac{1}{3})}{2}}$$

$$= \sqrt{\dfrac{2}{3}}$$

$$= \dfrac{\sqrt{6}}{3}$$

65. If $\cos A = -\dfrac{1}{3}$ and A is in QII, then

$$\sin A = \sqrt{1 - \cos^2 A}$$

$$= \sqrt{1 - \dfrac{1}{9}} = \sqrt{\dfrac{8}{9}} = \dfrac{2\sqrt{2}}{3}$$

Also, $\tan A = \dfrac{\sin A}{\cos A} = \dfrac{\frac{2\sqrt{2}}{3}}{\frac{1}{3}} = 2\sqrt{2}$

Therefore, $\tan 2A = \dfrac{2 \tan A}{1 - \tan^2 A} = \dfrac{2(2\sqrt{2})}{1 - (2\sqrt{2})^2} = \dfrac{4\sqrt{2}}{1 - 8} = -\dfrac{4\sqrt{2}}{7}$

69. $|\vec{g}| = \sqrt{(170)^2 + (28)^2 - 2(170)(28)\cos 68°}$

$$r = \sqrt{28{,}900 + 784 - 9520(0.3746)}$$

$$r = \sqrt{26{,}118}$$

$r = 160$ mph (to 2 significant digits)

$$\dfrac{\sin \theta}{28} = \dfrac{\sin 68°}{160}$$

$$\sin \theta = \dfrac{28 \sin 68°}{160}$$

$$\sin \theta = \dfrac{28(0.9272)}{160}$$

$$\sin \theta = 0.1623$$

$\theta = 9°$ (to the nearest degree)

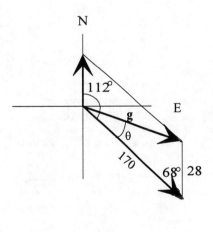

The true course is $112° - \theta = 112° - 9° = 103°$.

The ground speed is 160 miles per hour.

Problem Set 8.4

1. The 2 square roots will be

$$w_k = 4^{1/2}\left[\cos\frac{30° + 360°k}{2} + i\sin\frac{30° + 360°k}{2}\right] \text{ for } k = 0, 1$$

$$= 2[\cos(15° + 180°k) + i\sin(15° + 180°k)]$$

Replacing k with 0 and 1, we have:

when $k = 0$, $w_0 = 2(\cos 15° + i\sin 15°)$

when $k = 1$, $w_1 = 2(\cos 195° + i\sin 195°)$

5. The 2 square roots will be

$$w_k = 49^{1/2}\left[\cos\frac{180° + 360°\ k}{2} + i\sin\frac{180° + 360°\ k}{2}\right] \text{ for } k = 0, 1$$

$$= 7[\cos(90° + 180°\ k) + i\sin(90° + 180°\ k)]$$

Replacing k with 0 and 1, we have:

when $k = 0$, $w_0 = 7(\cos 90° + i\sin 90°)$

when $k = 1$, $w_1 = 7(\cos 270° + i\sin 270°)$

9. In trigonometric form, $4i = 4(\cos 90° + i\sin 90°)$.

The 2 square roots of $4i$ will be

$$w_k = 4^{1/2}\left[\cos\frac{90° + 360°k}{2} + i\sin\frac{90° + 360°k}{2}\right] \text{ for } k = 0, 1$$

$$= 2[\cos(45° + 180°k) + i\sin(45° + 180°k)]$$

Replacing k with 0 and 1, we have:

when $k = 0$, $w_0 = 2(\cos 45° + i\sin 45°)$

$$= 2\left(\frac{\sqrt{2}}{2} + i\cdot\frac{\sqrt{2}}{2}\right)$$

$$= \sqrt{2} + i\sqrt{2}$$

when $k = 1$, $w_1 = 2(\cos 225° + i\sin 225°)$

$$= 2\left[-\frac{\sqrt{2}}{2} + i\cdot\left(-\frac{\sqrt{2}}{2}\right)\right]$$

$$= -\sqrt{2} - i\sqrt{2}$$

13. In trigonometric form, $1 + i\sqrt{3} = 2(\cos 60° + i \sin 60°)$

The 2 square roots of $1 + i\sqrt{3}$ will be

$$w_k = 2^{1/2}\left[\cos\frac{60° + 360°k}{2} + i \sin\frac{60° + 360°k}{2}\right] \text{ for } k = 0, 1$$

$$= \sqrt{2}\left[\cos(30° + 180° k) + i \sin(30° + 180° k)\right]$$

Replacing k with 0 and 1, we have:

when $k = 0$, $w_0 = \sqrt{2}(\cos 30° + i \sin 30°)$

$$= \sqrt{2}\left(\frac{\sqrt{3}}{2} + i \cdot \frac{1}{2}\right)$$

$$= \frac{\sqrt{6}}{2} + \frac{i\sqrt{2}}{2}$$

when $k = 1$, $w_1 = \sqrt{2}(\cos 210° + i \sin 210°)$

$$= \sqrt{2}[-\frac{\sqrt{3}}{2} + i \cdot (-\frac{1}{2})]$$

$$= -\frac{\sqrt{6}}{2} - \frac{i\sqrt{2}}{2}$$

17. First, we write $4\sqrt{3} + 4i$ in trigonometric form:

$x = 4\sqrt{3}$, $y = 4$, and $r = \sqrt{(4\sqrt{3})^2 + 4^2} = 8$.

Also, $\tan\theta = \frac{4}{4\sqrt{3}} = \frac{1}{\sqrt{3}}$ and θ is in QI.

Therefore, $\theta = 30°$.

$4\sqrt{3} + 4i = 8(\cos 30° + i \sin 30°)$ in trigonometric form.

The 3 cube roots will be

$$w_k = 8^{1/3}\left[\cos\frac{30° + 360°k}{3} + i \sin\frac{30° + 360°k}{3}\right] \text{ for } k = 0, 1, 2$$

$$= 2[\cos(10° + 120°k) + i \sin(10° + 120° k)]$$

Replacing k with 0, 1, and 2, we have:

when $k = 0$, $w_0 = 2(\cos 10° + i \sin 10°)$

when $k = 1$, $w_1 = 2(\cos 130° + i \sin 130°)$

when $k = 2$, $w_2 = 2(\cos 250° + i \sin 250°)$

21. In trigonometric form, $64i = 64(\cos 90° + i \sin 90°)$

The 3 cube roots will be

$$w_k = 64^{1/3} \left[\cos \frac{90° + 360°k}{3} + i \sin \frac{90° + 360°k}{3} \right] \text{ for } k = 0, 1, 2$$

$$= 4[\cos(30° + 120°k) + i \sin(30° + 120°k)]$$

Replacing k with 0, 1, and 2, we have:

when $k = 0$, $w_0 = 4(\cos 30° + i \sin 30°)$

when $k = 1$, $w_1 = 4(\cos 150° + i \sin 150°)$

when $k = 2$, $w_2 = 4(\cos 270° + i \sin 270°)$

25. $x^4 - 16 = 0$

$x^4 = 16$ The solutions to this equation are the 4 fourth roots of 16.

In trigonometric form, $16 = 16(\cos 0° + i \sin 0°)$

The 4 fourth roots will be

$$w_k = 16^{1/4} \left[\cos \frac{0° + 360°k}{4} + i \sin \frac{0° + 360°k}{4} \right] \text{ for } k = 0, 1, 2, 3$$

$$= 2[\cos 90°k + i \sin 90°k]$$

Replacing k with 0, 1, 2, and 3, we have:

when $k = 0$, $w_0 = 2(\cos 0° + i \sin 0°)$

$$= 2(1 + i \cdot 0) = 2$$

when $k = 1$, $w_1 = 2(\cos 90° + i \sin 90°)$

$$= 2(0 + i \cdot 1) = 2i$$

when $k = 2$, $w_2 = 2(\cos 180° + i \sin 180°)$

$$= 2(-1 + i \cdot 0) = -2$$

when $k = 3$, $w_3 = 2(\cos 270° + i \sin 270°)$

$$= 2[0 + i(-1)] = -2i$$

29. The 5 fifth roots will be

$$w_k = (10^5)^{1/5} \left[\cos \frac{15° + 360°k}{5} + i \sin \frac{15° + 360°k}{5} \right] \text{ for } k = 0, 1, 2, 3, 4$$

$$= 10[\cos(3° + 72°k) + i \sin(3° + 72°k)]$$

Replacing k with 0, 1, 2, 3, and 4, we have:

when $k = 0$, $w_0 = 10(\cos 3° + i \sin 3°)$
$$= 10[0.999 + i(0.052)]$$
$$= 9.99 + 0.52i$$

when $k = 1$, $w_1 = 10(\cos 75° + i \sin 75°)$
$$= 10[0.259 + i(0.966)]$$
$$= 2.59 + 9.66i$$

when $k = 2$, $w_2 = 10(\cos 147° + i \sin 147°)$
$$= 10[-0.839 + i(0.545)]$$
$$= -8.39 + 5.45i$$

when $k = 3$, $w_3 = 10(\cos 219° + i \sin 219°)$
$$= 10[-0.777 + i(-0.629)]$$
$$= -7.77 - 6.29i$$

when $k = 4$, $w_4 = 10(\cos 291° + i \sin 291°)$
$$= 10[0.358 + i(-0.934)]$$
$$= 3.58 - 9.34i$$

33. Applying the quadratic formula, we have

$$x^2 = \frac{2 \pm \sqrt{(-2)^2 - 4(1)(4)}}{2(1)}$$

$$= \frac{2 \pm \sqrt{-12}}{2} = \frac{2 \pm 2i\sqrt{3}}{2} = 1 \pm i\sqrt{3}$$

That is, $x^2 = 1 + i\sqrt{3}$ or $x^2 = 1 - i\sqrt{3}$.

First, we find the 2 square roots of $1 + i\sqrt{3}$:

$1 + i\sqrt{3} = 2(\cos 60° + i \sin 60°)$ in trigonometric form

The 2 square roots will be

$$w_k = 2^{1/2}\left[\cos \frac{60° + 360°k}{2} + i \sin \frac{60° + 360°k}{2}\right] \text{ for } k = 0, 1$$

$$= \sqrt{2}[\cos(30° + 180°k) + \sin(30° + 180°k)]$$

Replacing k with 0 and 1, we have:

when $k = 0$, $w_0 = \sqrt{2}(\cos 30° + i \sin 30°)$

when $k = 1$, $w_1 = \sqrt{2}(\cos 210° + i \sin 210°)$

Continued on next page.

Second, we find the 2 square roots of $1 - i\sqrt{3}$:

$$1 - i\sqrt{3} = 2(\cos 300° + i \sin 300°) \text{ in trigonometric form}$$

The 2 square roots will be

$$w_k = 2^{1/2}\left[\cos\frac{300° + 360°k}{2} + i \sin\frac{300° + 360°k}{2}\right] \text{ for } k = 0, 1$$

$$= \sqrt{2}[\cos(150° + 180°k) + i \sin(150° + 180°k)]$$

Replacing k with 0 and 1, we have:

when $k = 0$, $w_0 = \sqrt{2}(\cos 150° + i \sin 150°)$

when $k = 1$, $w_1 = \sqrt{2}(\cos 330° + i \sin 330°)$

37. $y = -2\sin(-3x) = 2\sin 3x$ because sine is an odd function.

The graph is a sine curve with amplitude $= 2$ and period $= \frac{2\pi}{3}$.

41. The graph is a sine curve with

Amplitude $= 3$

Period $= \dfrac{2\pi}{\pi/3} = 6$

Phase shift $= \dfrac{\pi/3}{\pi/3} = 1$

45. $s = \dfrac{1}{2}(a + b + c)$ Formula for half-perimeter

    $= \dfrac{1}{2}(2.3 + 3.4 + 4.5)$ Substitute known values

    $= 5.1$ Simplify

$S = \sqrt{s(s - a)(s - b)(s - c)}$ Formula for area of triangle

    $= \sqrt{5.1(5.1 - 2.3)(5.1 - 3.4)(5.1 - 4.5)}$ Substitute known values

    $= \sqrt{5.1(2.8)(1.7)(0.6)}$ Simplify

    $= \sqrt{14.5656}$

    $= 3.8 \text{ ft}^2$ Round to 2 significant digits

Problem Set 8.5

13. All points of the form $(2, 60° + 360°k)$, where k is an integer, will name the point $(2, 60°)$.

For example, if $k = -1$, we have $(2, -300°)$.

Also, all points of the form $(-2, 240° + 360°k)$, where k is an integer, will name the point $(2, 60°)$. For example,

if $k = 0$, we have $(-2, 240°)$.

if $k = -1$, we have $(-2, -120°)$.

17. All points of the form $(-3, 30° + 360°k)$, where k is an integer, will name the point $(-3, 30°)$.

For example, if $k = -1$, we have $(-3, -330°)$.

Also, all points of the form $(3, 210° + 360°k)$, where k is an integer, will name the point $(-3, 30°)$. For example,

if $k = 0$, we have $(3, 210°)$.

if $k = -1$, we have $(3, -150°)$.

21.
$$x = r\cos\theta \qquad \text{and} \qquad y = r\sin\theta$$
$$= 3\cos 270° \qquad\qquad = 3\sin 270°$$
$$= 3(0) \qquad\qquad\qquad = 3(-1)$$
$$= 0 \qquad\qquad\qquad\qquad = -3$$
$$(3, 270°) = (0, -3)$$

25.
$$x = r\cos\theta \qquad \text{and} \qquad y = r\sin\theta$$
$$= -4\sqrt{3}\cos 30° \qquad\qquad = -4\sqrt{3}\sin 30°$$
$$= -4\sqrt{3}\left(\frac{\sqrt{3}}{2}\right) \qquad\qquad = -4\sqrt{3}\left(\frac{1}{2}\right)$$
$$= -6 \qquad\qquad\qquad\qquad = -2\sqrt{3}$$
$$(-4\sqrt{3}, 30°) = (-6, -2\sqrt{3})$$

29.
$$r = \sqrt{x^2 + y^2} \qquad \text{and} \qquad \tan\theta = \frac{y}{x}$$
$$= \sqrt{(-2\sqrt{3})^2 + (2)^2} \qquad\qquad = \frac{2}{-2\sqrt{3}}$$
$$= \sqrt{12 + 4} \qquad\qquad\qquad = -\frac{1}{\sqrt{3}}$$
$$= 4 \qquad\qquad\qquad\qquad \theta = 150° \text{ or } 330°$$

Since $(-2\sqrt{3}, 2)$ is in QII, one solution is $(4, 150°)$.

33. $r = \sqrt{x^2 + y^2}$ and $\tan\theta = \dfrac{y}{x}$

$\qquad = \sqrt{(-\sqrt{3})^2 + (-1)^2}$ $= \dfrac{-1}{-\sqrt{3}}$

$\qquad = \sqrt{3 + 1}$ $= \dfrac{1}{\sqrt{3}}$

$\qquad = 2$ $\theta = 30° \text{ or } 210°$

Since $(-\sqrt{3}, -1)$ is in QIII, one solution is $(2, 210°)$.

37. $r = \sqrt{x^2 + y^2}$ and $\tan\theta = \dfrac{y}{x}$

$\qquad = \sqrt{(-1)^2 + (2)^2}$ $= \dfrac{2}{-1}$

$\qquad = \sqrt{1 + 4}$ $\theta = 116.6° \text{ or } 296.6°$

$\qquad = \sqrt{5}$

Since $(-1, 2)$ is in QII, one solution is $(\sqrt{5}, 116.6°)$.

41. $\qquad r^2 = 9$

$\quad x^2 + y^2 = 9$ Substitute $x^2 + y^2$ for r^2

45. $\qquad r^2 = 4\sin 2\theta$

$\qquad r^2 = 4(2\sin\theta\cos\theta)$ Double-angle identity

$\qquad r^2 = 8\left(\dfrac{y}{r}\right)\left(\dfrac{x}{r}\right)$ Substitute $\dfrac{y}{r}$ for $\sin\theta$ and $\dfrac{x}{r}$ for $\cos\theta$

$\qquad r^2 = \dfrac{8xy}{r^2}$ Multiply

$\qquad r^4 = 8xy$ Multiply both sides by r^2

$\quad (x^2 + y^2)^2 = 8xy$ Substitute $x^2 + y^2$ for r^2

49. $\qquad x - y = 5$

$\quad r\cos\theta - r\sin\theta = 5$ Substitute $r\cos\theta$ for x and $r\sin\theta$ for y

$\quad r(\cos\theta - \sin\theta) = 5$ Factor out r

53. $x^2 + y^2 = 6x$

$\qquad r^2 = 6r\cos\theta$ Substitute r^2 for $x^2 + y^2$ and $r\cos\theta$ for x

$\qquad r = 6\cos\theta$ Divide both sides by r

57. The graph is a sine curve with amplitude of 6.

61. The graph is a sine curve with amplitude of 2, period of 2π, and a vertical translation of 4.

Problem Set 8.6

1.

θ	$r = 6\cos\theta$	(r, θ)
$0°$	$r = 6\cos 0°$	$(6, 0°)$
$45°$	$r = 6\cos 45° = 4.2$	$(4.2, 270°)$
$90°$	$r = 6\cos 90° = 0$	$(0, 90°)$
$135°$	$r = 6\cos 135° = -4.2$	$(-4.2, 135°)$
$180°$	$r = 6\cos 180° = -6$	$(-6, 180°)$
$225°$	$r = 6\cos 225° = -4.2$	$(-4.2, 225°)$
$270°$	$r = 6\cos 270° = 0$	$(0, 270°)$
$315°$	$r = 6\cos 315° = 4.2$	$(4.2, 315°)$

5.

$$r = 3$$

$$\pm\sqrt{x^2 + y^2} = 3 \qquad \text{Substitute } \pm\sqrt{x^2 + y^2} \text{ for } r$$

$$x^2 + y^2 = 9 \qquad \text{Square both sides}$$

The graph is a circle with center at the origin or pole and radius of 3.

9. To graph by hand, first we sketch $y = 3\sin x$:

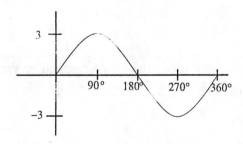

Next, we note the relationship between variations in x (or θ) and the corresponding variations in y (or r).

Variations in x (or θ)	Corresponding Variations in y (or r)
$0°$ to $90°$	0 to 3
$90°$ to $180°$	3 to 0
$180°$ to $270°$	0 to -3
$270°$ to $360°$	-3 to 0

Then, we sketch the graph using this relationship.

To graph on a graphing calculator:

1. Set $\boxed{\text{MODE}}$ to $\boxed{\text{DEGREE}}$ and $\boxed{\text{POL}}$

2. Press $\boxed{\text{Y} =}$ and enter R1 = 3 $\boxed{\sin}$ $\boxed{(}$ $\boxed{\theta}$ $\boxed{)}$.

3. Press $\boxed{\text{WINDOW}}$ and enter the following settings:

$$\theta \min = 0$$
$$\theta \max = 360$$
$$\theta \text{ step} = 7.5$$
$$X \min = -5.5$$
$$X \max = 5.5$$
$$X \text{scl} = 1$$
$$Y \min = -3.8$$
$$Y \max = 3.8$$
$$Y \text{scl} = 1$$

4. Then, press $\boxed{\text{GRAPH}}$.

13. First we sketch $y = 2 + 4 \cos x$:

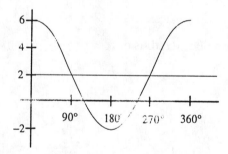

Next, we note the relationship between variations in x (or θ) and the corresponding variations in y (or r).

Variations in x (or θ)	Corresponding Variations in y (or r)
0° to 90°	6 to 2
90° to 180°	2 to −2
180° to 270°	−2 to 2
270° to 360°	2 to 6

Then, we sketch the graph using this relationship.

17. $r^2 = 9 \sin 2\theta$ is a lemniscate or two-leaved rose. (See Figure 13 in textbook.) The endpoints of the leaves are $(\sqrt{a}, 45°) = (3, 45°)$ and $(-\sqrt{a}, 45°) = (-3, 45°)$.

21. $r = 4 \cos 3\theta$ is a three-leaved rose. (See Figure 12 in textbook). The endpoints of the leaves are $(a, 0°) = (4, 0°)$, $(a, 120°) = (4, 120°)$, and $(a, 240°) = (4, 240°)$.

25. $x^2 + y^2 = 6x$

$\qquad r^2 = 6r \cos\theta \qquad$ Substitute r^2 for $x^2 + y^2$ and $r \cos\theta$ for x

$\qquad r = 6 \cos\theta \qquad$ Divide both sides by r

The graph is a circle with center at $(3, 0°)$ and radius $= 3$.

29. $r(2 \cos\theta + 3 \sin\theta) = 6$

$\qquad 2r \cos\theta + 3r \sin\theta = 6 \qquad$ Multiply

$\qquad\qquad 2x + 3y = 6 \qquad$ Substitute x for $r \cos\theta$ and y for $r \sin\theta$

The graph is a line through $(0, 2)$ and $(3, 0)$ in rectangular coordinates, or $(3, 0°)$ and $(2, 90°)$ in polar coordinates.

33. $\qquad\qquad r = 4 \sin\theta$

$\qquad\qquad r^2 = 4r \sin\theta \qquad$ Multiply both sides by r

$\qquad x^2 + y^2 = 4y \qquad$ Substitute $x^2 + y^2$ for r^2 and y for $r \sin\theta$

$\qquad x^2 + y^2 - 4y = 0 \qquad$ Subtract $4y$ from both sides

$\qquad x^2 + y^2 - 4y + 4 = 4 \qquad$ Complete the square by adding 4 to both sides

$\qquad x^2 + (y - 2)^2 = 4 \qquad$ Factor

The graph is a circle with center at $(0, 2)$ or $(2, 90°)$ and radius $= 2$.

37. Let $y_1 = \sin x$ (the basic sine curve) and $y_2 = -\cos x$ (a cosine curve reflected about the x-axis). Graph y_1, y_2, and $y = y_1 + y_2$ on the same coordinate axes.

41. Let $y_1 = 3 \sin x$ (a sine curve with amplitude of 3) and $y_2 = \cos 2x$ (a cosine curve with period of $\dfrac{2\pi}{2}$ or π). Graph y_1, y_2, and $y = y_1 + y_2$ on the same coordinate axes.

Chapter 8 Test

1. $\sqrt{-25} = \sqrt{25}\, i = 5i$

2. $\sqrt{-12} = \sqrt{12}\, i$
   $\qquad = 2i\sqrt{3}$

3. $7x = 14$ and $-6 = -3y$
   $\quad x = 2 \qquad\qquad y = 2$

4. $\qquad x^2 - 3x = 10$ and $16 = 8y$
   $\quad x^2 - 3x - 10 = 0 \qquad\quad y = 2$
   $\quad (x - 5)(x + 2) = 0$
   $\quad x - 5 = 0$ or $x + 2 = 0$
   $\qquad x = 5 \qquad\qquad x = -2$

5. $(6 - 3i) + [(4 - 2i) - (3 + i)] = 6 - 3i + 4 - 2i - 3 - i$
   $\qquad\qquad\qquad\qquad\qquad = (6 + 4 - 3) + (-3i - 2i - i)$
   $\qquad\qquad\qquad\qquad\qquad = 7 - 6i$

6. $(7 + 3i) - [(2 + i) - (3 - 4i)] = 7 + 3i - [2 + i - 3 + 4i]$
   $\qquad\qquad\qquad\qquad\qquad = 7 + 3i - (-1 + 5i)$
   $\qquad\qquad\qquad\qquad\qquad = 7 + 3i + 1 - 5i$
   $\qquad\qquad\qquad\qquad\qquad = 8 - 2i$

7. $i^{16} = (i^4)^4$
   $\quad\ = 1^4$
   $\quad\ = 1$

8. $i^{17} = (i^4)^4\, i$
   $\quad\ = 1 \cdot i$
   $\quad\ = i$

9. $(8 + 5i)(8 - 5i) = 64 + 40i - 40i - 25i^2$
   $\qquad\qquad\qquad = 64 - 25(-1)$
   $\qquad\qquad\qquad = 64 + 25$
   $\qquad\qquad\qquad = 89$

10. $(3 + 5i)^2 = 9 + 30i + 25i^2$

$$= 9 + 30i + 25(-1)$$

$$= 9 + 30i - 25$$

$$= -16 + 30i$$

11. $\dfrac{5 - 4i}{2i} \cdot \dfrac{i}{i} = \dfrac{5i - 4i^2}{2i^2}$

$$= \dfrac{5i - 4(-1)}{2(-1)}$$

$$= \dfrac{4 + 5i}{-2}$$

$$= -2 - \dfrac{5}{2}i$$

12. $\dfrac{6 + 5i}{6 - 5i} \cdot \dfrac{6 + 5i}{6 + 5i} = \dfrac{36 + 60i + 25i^2}{36 - 25i^2}$

$$= \dfrac{36 + 60i + 25(-1)}{36 - 25(-1)}$$

$$= \dfrac{36 + 60i - 25}{36 + 25}$$

$$= \dfrac{11 + 60i}{61}$$

$$= \dfrac{11}{61} + \dfrac{60}{61}i$$

13. **a.** $|3 + 4i| = \sqrt{(3)^3 + (4)^2}$

$$= \sqrt{25}$$

$$= 5$$

b. $-(3 + 4i) = -3 - 4i$

c. The conjugate of $3 + 4i$ is $3 - 4i$.

14. **a.** $|3 - 4i| = \sqrt{3^2 + (-4)^2}$

$$= \sqrt{9 + 16}$$

$$= \sqrt{25}$$

$$= 5$$

Continued on next page.

b. $-(3 - 4i) = -3 + 4i$

c. The conjugate of $3 - 4i$ is $3 + 4i$.

15. a. $|8i| = \sqrt{0^2 + 8^2}$

$= \sqrt{64}$

$= 8$

b. $-(8i) = -8i$

c. The conjugate of $0 + 8i$ is $0 - 8i$ or $-8i$.

16. a. $|-4| = 4$

b. $-(-4) = 4$

c. The conjugate of $4 + 0i$ is $4 - 0i$ or 4.

17. $8(\cos 330° + i \sin 330°) = 8[\dfrac{\sqrt{3}}{2} + i(-\dfrac{1}{2})]$

$= 4\sqrt{3} - 4i$

18. $2(\cos 135° + i \sin 135°) = 2[-\dfrac{\sqrt{2}}{2} + i(\dfrac{\sqrt{2}}{2})]$

$= -\sqrt{2} + i\sqrt{2}$

19. $x = 2$ and $y = 2$

$r = \sqrt{2^2 + 2^2}$ $\qquad \tan\theta = \dfrac{2}{2}$ and θ is in QI

$= \sqrt{4 + 4}$ $\qquad\qquad = 1$

$= \sqrt{8}$ $\qquad\qquad \theta = 45°$

$= 2\sqrt{2}$

Therefore, $2 + 2i = 2\sqrt{2}(\cos 45° + i \sin 45°)$

20. $x = -\sqrt{3}$ and $y = 1$

$r = \sqrt{(-\sqrt{3})^2 + 1^2}$ $\qquad \tan\theta = \dfrac{1}{-\sqrt{3}}$ and θ is in QII

$= \sqrt{3 + 1}$ $\qquad\qquad \theta = 150°$

$= 2$

Therefore, $-\sqrt{3} + i = 2(\cos 150° + i \sin 150°)$

21. $5i$ lies on the positive y-axis.

Therefore, $r = 5$ and $\theta = 90°$.

$5i = 5(\cos 90° + i \sin 90°)$

22. -3 lies on the negative x-axis.

Therefore, $r = 3$ and $\theta = 180°$.

$-3 = 3(\cos 180° + i \sin 180°)$

23. $5(\cos 25° + i \sin 25°) \cdot 3(\cos 40° + i \sin 40°)$

$\qquad = 5 \cdot 3[\cos(25° + 40°) + i \sin(25° + 40°)]$

$\qquad = 15(\cos 65° + i \sin 65°)$

24. $\dfrac{10(\cos 50° + i \sin 50°)}{2(\cos 20° + i \sin 20°)} = \dfrac{10}{2}[\cos(50° - 20°) + i \sin(50° - 20°)]$

$\qquad\qquad\qquad = 5(\cos 30° + i \sin 30°)$

25. $[2(\cos 10° + i \sin 10°)]^5 = 2^5[\cos(5 \cdot 10°) + i \sin(5 \cdot 10°)]$

$\qquad\qquad\qquad = 32(\cos 50° + i \sin 50°)$

26. $[3(\cos 20° + i \sin 20°)]^4 = 3^4[\cos(4 \cdot 20°) + i \sin(4 \cdot 20°)]$

$\qquad\qquad\qquad = 81(\cos 80° + i \sin 80°)$

27. The 2 square roots will be

$$w_k = 49^{1/2}\left[\cos\frac{50° + 360°k}{2} + i \sin\frac{50° + 360°\,k}{2}\right] \text{ for } k = 0, 1$$

$$= 7[\cos(25° + 180°k) + i \sin(25° + 180°k)]$$

For $k = 0$, $w_0 = 7(\cos 25° + i \sin 25°)$

For $k = 1$, $w_1 = 7(\cos 205° + i \sin 205°)$

28. First, we write $2 + 2i\sqrt{3}$ in trigonometric form:

$2 + 2i\sqrt{3} = 4(\cos 60° + i \sin 60°)$

The 4 fourth roots will be

Continued on next page.

$$w_k = 4^{1/4}\left[\cos\frac{60° + 360° \, k}{4} + i \sin\left(\frac{60° + 360°}{4}\right)\right] \text{ for } k = 0, 1, 2, 3$$

$$= \sqrt{2}[\cos(15° + 90° \, k) + i \sin(15° + 90° \, k)]$$

For $k = 0$, $w_0 = \sqrt{2}(\cos 15° + i \sin 15°)$

For $k = 1$, $w_1 = \sqrt{2}(\cos 105° + i \sin 105°)$

For $k = 2$, $w_2 = \sqrt{2}(\cos 195° + i \sin 195°)$

For $k = 3$, $w_3 = \sqrt{2}(\cos 285° + i \sin 285°)$

29. Using the quadratic formula, we have

$$x^2 = \frac{2\sqrt{3} \pm \sqrt{12 - 4(1)(4)}}{2(1)}$$

$$= \frac{2\sqrt{3} \pm \sqrt{-4}}{2}$$

$$= \frac{2\sqrt{3} \pm 2i}{2}$$

$$= \sqrt{3} \pm i$$

Therefore, $x^2 = \sqrt{3} + i$ or $x^2 = \sqrt{3} - i$.

First, we will find the 2 square roots of $\sqrt{3} + i$:

$\sqrt{3} + i = 2(\cos 30° + i \sin 30°)$ in trigonometric form

The 2 square roots will be

$$w_k = 2^{1/2}\left[\cos\frac{30° + 360°k}{2} + i \sin\frac{30° + 360°k}{2}\right] \text{ for } k = 0, 1$$

$$= \sqrt{2}[\cos(15° + 180°k) + i \sin(15° + 180°k)]$$

For $k = 0$, $w_0 = \sqrt{2}(\cos 15° + i \sin 15°)$

For $k = 1$, $w_1 = \sqrt{2}(\cos 195° + i \sin 195°)$

Next, we will find the 2 square roots of $\sqrt{3} - i$:

$\sqrt{3} - i = 2(\cos 330° + i \sin 330°)$ in trigonometric form

The 2 square roots will be

$$w_k = 2^{1/2}\left[\cos\frac{330° + 360°k}{2} + i \sin\frac{330° + 360°k}{2}\right] \text{ for } k = 0, 1$$

$$= \sqrt{2}[\cos(165° + 180°k) + i \sin(165° + 180°k)]$$

For $k = 0$, $w_0 = \sqrt{2}(\cos 165° + i \sin 165°)$

For $k = 1$, $w_1 = \sqrt{2}(\cos 345° + i \sin 345°)$

30. First we write -1 in trigonometric form:

$$-1 = 1(\cos 180° + i \sin 180°)$$

The 3 cube roots will be

$$w_k = 1^{1/3} \left[\cos \frac{180° + 360° \, k}{3} + i \sin \frac{180° + 360° \, k}{3} \right] \text{ for } k = 0, 1, 2$$

$$= 1[\cos(60° + 120° \, k) + i \sin(60° + 120° \, k)]$$

For $k = 0$, $w_0 = \cos 60° + i \sin 60°$

For $k = 1$, $w_1 = \cos 180° + i \sin 180°$

For $k = 2$, $w_2 = \cos 300° + i \sin 300°$

31. All points of the form $(4, 225° + 360° \, k)$, where k is an integer, will name the point $(4, 225°)$. For example, if $k = -1$, we have $(4, -135°)$.

Also, all points of the form $(-4, 45° + 360° \, k)$, where k is an integer, will name the point $(4, 225°)$. For example, if $k = 0$, we have $(-4, 45°)$

Since $r = 4$ and $\theta = 225°$,

$$x = r \cos \theta \qquad \text{and} \qquad y = r \sin \theta$$

$$= 4 \cos 225° \qquad\qquad\qquad = 4 \sin 225°$$

$$= 4\left(-\frac{\sqrt{2}}{2}\right) \qquad\qquad\qquad = 4\left(-\frac{\sqrt{2}}{2}\right)$$

$$= -2\sqrt{2} \qquad\qquad\qquad\qquad = -2\sqrt{2}$$

Therefore, $(4, 225°) = (-2\sqrt{2}, -2\sqrt{2})$.

32. All points of the form $(-6, 60° + 360° \, k)$, where k is an integer, will name the point $(-6, 60°)$. For example, if $k = -1$, we have $(-6, -300°)$.

Also, all points of the form $(6, 240° + 360° \, k)$, where k is an integer, will name the point $(-6, 60°)$. For example, if $k = -1$, we have $(6, -120°)$ and if $k = 0$, we have $(6, 240°)$.

Continued on next page.

Since $r = -6$ and $\theta = 60°$,

$$x = r \cos \theta \qquad \text{and} \qquad y = r \sin \theta$$
$$= -6 \cos 60° \qquad\qquad\qquad = -6 \sin 60°$$
$$= -6\left(\frac{1}{2}\right) \qquad\qquad\qquad = -6\left(\frac{\sqrt{3}}{2}\right)$$
$$= -3 \qquad\qquad\qquad\qquad = -3\sqrt{3}$$

Therefore, $(-6, 60°) = (-3, -3\sqrt{3})$.

33. $r = \sqrt{x^2 + y^2} \qquad \text{and} \qquad \tan \theta = \dfrac{y}{x}$

$\qquad = \sqrt{(-3)^2 + (3)^2} \qquad \tan \theta = \dfrac{3}{-3} = -1$

$\qquad = \sqrt{18} \qquad\qquad\qquad \theta = 135° \text{ or } 315°$

$\qquad = 3\sqrt{2}$

Since $(-3, 3)$ is in QII, one solution is $(3\sqrt{2}, 135°)$.

34. Since $(0, 5)$ lies on the positive y-axis, $r = 5$ and $\theta = 90°$. Therefore, $(0, 5) = (5, 90°)$.

35. $\qquad\qquad r = 6 \sin \theta$

$\qquad\qquad r^2 = 6\, r \sin \theta \qquad\qquad$ Multiply both sides by r

$\qquad x^2 + y^2 = 6y \qquad\qquad$ Substitute $x^2 + y^2$ for r^2 and y for $r \sin \theta$

36. $\qquad\qquad r = \sin 2\theta$

$\qquad\qquad r = 2 \sin \theta \cos \theta \qquad\qquad$ Double-angle formula

$\qquad\qquad r = 2\left(\dfrac{y}{r}\right)\left(\dfrac{x}{r}\right) \qquad\qquad$ Substitute $\dfrac{y}{r}$ for $\sin \theta$ and $\dfrac{x}{r}$ for $\cos \theta$

$\qquad\qquad r = \dfrac{2xy}{r^2} \qquad\qquad$ Simplify

$\qquad\qquad r^3 = 2xy \qquad\qquad$ Multiply both sides by r^2

$\qquad\qquad (r^2)^{2/3} = 2xy \qquad\qquad$ Rewrite in terms of r^2

$\qquad (x^2 + y^2)^{2/3} = 2xy \qquad\qquad$ Substitute $x^2 + y^2$ for r^2

37. $\qquad\qquad x + y = 2$

$\qquad r \cos \theta + r \sin \theta = 2 \qquad\qquad$ Substitute $r \cos \theta$ for x and $r \sin \theta$ for y

$\qquad r(\cos \theta + \sin \theta) = 2 \qquad\qquad$ Factor out r

38. $x^2 + y^2 = 8y$

$\qquad r^2 = 8\,r\sin\theta \qquad\qquad$ Substitute r^2 for $x^2 + y^2$ and $r\sin\theta$ for y

$\qquad r = 8\sin\theta \qquad\qquad$ Divide both sides by r

39. The graph of $r = 4$ is a circle with center at the pole and the radius of 4.

40. The graph of $\theta = 45°$ is the straight line through the pole where θ is $45°$. This is the same as the line $y = x$ in rectangular coordinates.

41. First, we sketch $y = 4 + 2\cos x$:

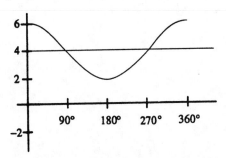

Next, we note the relationship between variations in x (or θ), and the corresponding variations in y (or r).

Variations in x (or θ)	Corresponding Variations in y (or r)
0° to 90°	6 to 4
90° to 180°	4 to 2
180° to 270°	2 to 4
270° to 360°	4 to 6

Then, we sketch the graph using this relationship.

42. The graph of $r = \sin 2\theta$ is a 4-leaved rose. (See Figure 11 in textbook). The endpoints of the leaves are $(a, 45°) = (1, 45°)$, $(a, 135°) = (1, 135°)$, $(a, 225°) = (1, 225°)$, and $(a, 315°) = (1, 315°)$.

APPENDIX A Logarithms

Problem Set A.1

1. $g(0) = (\frac{1}{2})^0$

$\qquad = 1$

5. $f(-3) = (3)^{-3}$

$\qquad = \dfrac{1}{3^3}$

$\qquad = \dfrac{1}{27}$

9. If we let $x = -1$, the equation $y = 4^x$ becomes:

$$y = 4^{-1}$$

$$= \frac{1}{4}$$

This gives us $(-1, \dfrac{1}{4})$ as a point on the curve.

If we let $x = 0$, the equation $y = 4^x$ becomes:

$$y = 4^0$$

$$y = 1$$

This gives us $(0, 1)$ as a point on the curve.

If we let $x = 1$, the equation $y = 4^x$ becomes:

$$y = 4^1$$

$$= 4$$

This gives us $(1, 4)$ as a point on the curve.

Graph the points $(-1, \frac{1}{4})$, $(0, 1)$, and $(1, 4)$ and then draw a smooth curve through these points.

Also remember that y can never be equal to zero and therefore, the x-axis is a horizontal asymptote.

13. If we let $x = -1$, the equation $y = 2^{x+1}$ becomes:

$$y = 2^{-1+1}$$

$$= 2^0$$

$$= 1$$

This gives us $(-1, 1)$ as a point on the curve.

If we let $x = 0$, the equation $y = 2^{x+1}$ becomes:

$$y = 2^{0+1}$$
$$= 2^1$$
$$= 2$$

This gives us $(0, 2)$ as a point on the curve.

If we let $x = 1$, the equation $y = 2^{x+1}$ becomes:

$$y = 2^{1+1}$$
$$= 2^2$$
$$= 4$$

This gives us $(1, 4)$ as a point on the curve.

Graph the points $(-1, 1)$, $(0, 2)$, and $(1, 4)$ and then draw the smooth curve through them. (Remember that the x-axis is a horizontal asymptote.)

17. The graph of $y = 2x$ is a straight line.

If we let $x = 0$, then $y = 2(0) = 0$.

If we let $x = 2$, then $y = 2(2) = 4$

We draw the straight line through $(0, 0)$ and $(2, 4)$.

The graph of $y = x^2$ is a parabola.

If we let $x = 0$, then $y = 0^2 = 0$.

If we let $x = 1$, then $y = 1^2 = 1$.

If we let $x = 2$, then $y = 2^2 = 4$.

We draw the curve through $(0, 0)$, $(1, 1)$, and $(2, 4)$.

The graph of $y = 2^x$ is an exponential function.

If we let $x = 0$, then $y = 2^0 = 1$.

If we let $x = 1$, then $y = 2^1 = 2$.

If we let $x = 2$, then $y = 2^2 = 4$.

We draw the curve through $(0, 1)$, $(1, 2)$, and $(2, 4)$.

21. For the graph of $y = 3^x$:

If we let $x = -1$, then $y = 3^{-1} = \dfrac{1}{3}$.

If we let $x = 0$, then $y = 3^0 = 1$.

If we let $x = 1$, then $y = 3^1 = 3$.

We draw the curve through $(-1, \dfrac{1}{3})$, $(0, 1)$, and $(1, 3)$.

(Remember the x-axis is a horizontal asymptote.)

For the graph of the inverse, $x = 3^y$, the x- and y-coordinates are interchanged. We draw the curve through $(\frac{1}{3}, -1)$, $(1, 0)$, and $(3, 1)$. (Since x and y are interchanged, the y-axis will be a vertical asymptote.)

25. $h = \left(\dfrac{2}{3}\right)^n \cdot (6)$ where the initial height is 6 feet.

$$= 6 \cdot \left(\dfrac{2}{3}\right)^n$$

If $n = 5$, $h = 6 \cdot \left(\dfrac{2}{3}\right)^5$

$$= 0.79 \text{ feet}$$

29. Graph $y = .85^x$ on your graphing calculator.

Set the window for X between 0 and 10 with the Xscl $= 1$ and for Y between 0 and 1 with the Yscl $= .25$. Use the trace feature to find x when $y = .5$. The answer is approximately 4.3 days.

33. $A = P\left(1 + \dfrac{r}{n}\right)^{nt}$ where $P = 200$, $r = .08$, $n = 2$, and $t = 10$.

$$= 200\left(1 + \dfrac{.08}{2}\right)^{2 \cdot 10}$$

$$= 200(1.04)^{20}$$

$$= \$438.22$$

Problem Set A.2

17. $\log_3 x = 2$

$\quad\quad x = 3^2$ Change to exponential form

$\quad\quad x = 9$ Simplify

21. $\log_x 4 = 2$

$\quad\quad x^2 = 4 \quad\quad\quad$ Change to exponential form

$\quad\quad x = \pm 2 \quad\quad\quad$ Take square root of both sides

$\quad\quad x = 2 \quad\quad\quad\quad$ x must be positive

25. $y = \log_{1/3} x$

$\quad\quad x = \left(\dfrac{1}{3}\right)^y \quad\quad$ Change to exponential form

x	9	3	1	$\frac{1}{3}$	$\frac{1}{9}$
y	-2	-1	0	1	2

Graph these ordered pairs and join them with a smooth curve.

29. $\quad y = \log_{25} 125$

$\quad\quad 25^y = 125 \quad\quad\quad$ Change to exponential form

$\quad\quad (5^2)^y = 5^3 \quad\quad\quad$ Write 25 and 125 as powers of 5

$\quad\quad 2y = 3 \quad\quad\quad\quad$ If $b^m = b^n$, then $m = n$

$\quad\quad y = \dfrac{3}{2} \quad\quad\quad\quad$ Divide both sides by 2

33. $\log_3 3 = 1 \quad$ (See example 7 in textbook.)

37. $\log_3 (\log_6 6) = \log_3 1 \quad\quad \log_b b = 1$

$\quad\quad\quad\quad\quad\quad\quad = 0 \quad\quad\quad\quad \log_b 1 = 0$

41. $\quad \text{pH} = -\log_{10} [\text{H}^+]$

$\quad\quad\quad = -\log_{10}(10^{-7}) \quad\quad$ Substitute given values

$\quad\quad\quad = -(-7) \quad\quad\quad\quad\quad \log_b b^x = x$

$\quad\quad\quad = 7 \quad\quad\quad\quad\quad\quad\quad\quad$ Simplify

45. $M = \log_{10} T \quad\quad\quad$ Formula given in textbook

$\quad\quad = \log_{10} 100 \quad\quad\quad$ Substitute given values

$\quad\quad = \log_{10} 10^2 \quad\quad\quad$ Write 100 as a power of 10

$\quad\quad = 2 \quad\quad\quad\quad\quad\quad\quad \log_b b^x = x$

Problem Set A.3

1. $\log_3 4x = \log_3 4 + \log x$ Property 1

5. $\log_2 y^5 = 5 \log_2 y$ Property 3

9. $\log_6 x^2 y^3 = \log_6 x^2 + \log_6 y^3$ Property 1
$$= 2 \log_6 x + 3 \log_6 y \quad \text{Property 3}$$

13. $\log_b \dfrac{xy}{z} = \log_b (xy) - \log_b z$ Property 2
$$= \log_b x + \log_b y - \log_b z \quad \text{Property 1}$$

17. $\log_{10} \dfrac{x^2 y}{\sqrt{z}} = \log_{10} (x^2 y) - \log_{10} z^{1/2}$ Property 2
$$= \log_{10} x^2 + \log_{10} y - \log_{10} z^{1/3} \quad \text{Property 1}$$
$$= 2 \log_{10} x + \log_{10} y - \tfrac{1}{2} \log_{10} z \quad \text{Property 3}$$

21. $\log_b \sqrt[3]{\dfrac{x^2 y}{z^4}} = \log_b \left(\dfrac{x^2 y}{z^4} \right)^{1/3}$ Write as exponent
$$= \log_b \dfrac{x^{2/3} y^{1/3}}{z^{4/3}} \quad \text{Property of exponents}$$
$$= \log_b (x^{2/3} y^{1/3}) - \log_b z^{4/3} \quad \text{Property 2}$$
$$= \log_b x^{2/3} + \log_b y^{1/3} - \log_b z^{4/3} \quad \text{Property 1}$$
$$= \dfrac{2}{3} \log_b x + \dfrac{1}{3} \log_b y - \dfrac{4}{3} \log_b z \quad \text{Property 3}$$

25. $2 \log_3 x - 3 \log_3 y = \log_3 x^2 - \log_3 y^3$ Property 3
$$= \log_3 \dfrac{x^2}{y^3} \quad \text{Property 2}$$

29. $3 \log_2 x + \dfrac{1}{2} \log_2 y - \log_2 z = \log_2 x^3 + \log_2 y^{1/2} - \log_2 z$ Property 3
$$= \log_2 x^3 y^{1/2} - \log_2 z \quad \text{Property 1}$$
$$= \log_2 \dfrac{x^3 y^{1/2}}{z} \ \text{ or } \ \log_2 \dfrac{x^3 \sqrt{y}}{z} \quad \text{Property 2}$$

33. $\dfrac{3}{2} \log_{10} x - \dfrac{3}{4} \log_{10} y - \dfrac{4}{5} \log_{10} z$

$$= \log_{10} x^{3/2} - \log_{10} y^{3/4} - \log_{10} z^{4/5} \qquad \text{Property 3}$$

$$= \log_{10} x^{3/2} - \left(\log_{10} y^{3/4} + \log_{10} z^{4/5} \right) \qquad \text{Introduce parentheses}$$

$$= \log_{10} x^{3/2} - \log_{10} (y^{3/4} z^{4/5}) \qquad \text{Property 1}$$

$$= \log_{10} \left(\frac{x^{3/2}}{y^{3/4} z^{4/5}} \right) \qquad \text{Property 2}$$

37. $\log_3 x - \log_3 2 = 2$

$$\log_3 \frac{x}{2} = 2 \qquad \text{Property 2}$$

$$\frac{x}{2} = 3^2 \qquad \text{Change to exponential form}$$

$$\frac{x}{2} = 9 \qquad \text{Simplify}$$

$$x = 18 \qquad \text{Multiply both sides by 2 and check}$$

41. $\log_3(x + 3) - \log_3(x - 1) = 1$

$$\log_3 \left(\frac{x+3}{x-1} \right) = 1 \qquad \text{Property 2}$$

$$\frac{x+3}{x-1} = 3^1 \qquad \text{Change to exponential form}$$

$$x + 3 = 3x - 3 \qquad \text{Multiply both sides by } x - 1$$

$$-2x = -6 \qquad \text{Solve the linear equation and check}$$

$$x = 3$$

45. $\log_8 x + \log_8 (x - 3) = \dfrac{2}{3}$

$$\log_8 x(x - 3) = \frac{2}{3} \qquad \text{Property 1}$$

$$x(x - 3) = 8^{2/3} \qquad \text{Change to exponential form}$$

$$x^2 - 3x = 4 \qquad \text{Simplify}$$

$$x^2 - 3x - 4 = 0 \qquad \text{Put in standard form}$$

$$(x - 4)(x + 1) = 0 \qquad \text{Factor left side}$$

$$x - 4 = 0 \quad \text{or} \quad x + 1 = 0 \qquad \text{Set each factor} = 0$$

$$x = 4 \qquad\qquad x = -1 \qquad \text{Solve and check}$$

Solution is 4 $\qquad\qquad\qquad$ -1 does not check

49. $M = 0.21(\log_{10} a - \log_{10} b)$

$\quad\quad = 0.21(\log_{10} 1 - \log_{10} 10^{-12})$ Substitute given values

$\quad\quad = 0.21[0 - (-12)]$ $\log_b 1 = 0$ and $\log_b b^x = x$

$\quad\quad = 0.21(12)$ Simplify

$\quad\quad = 2.52$ Multiply

$M = 0.21 \log_{10} \left(\dfrac{a}{b}\right)$

$\quad\quad = 0.21 \log_{10} \left(\dfrac{1}{10^{-12}}\right)$ Substitute given values

$\quad\quad = 0.21 \log_{10} 10^{12}$ Simplify

$\quad\quad = 0.21(12)$ $\log_b b^x = x$

$\quad\quad = 2.52$ Multiply

53. $\log_{10} A = \log_{10} 100(1.06)^t$

$\quad\quad = \log_{10} 100 + \log_{10} (1.06)^t$ Property 1

$\quad\quad = \log_{10} (10^2) + t \log_{10} (1.06)$ Property 3

$\quad\quad = 2 + t \log_{10} (1.06)$ $\log_b b^x = x$

Problem Set A.4

1. Scientific Calculator: 378 $\boxed{\text{log}}$

Graphing Calculator: $\boxed{\text{log}}$ $\boxed{(}$ 378 $\boxed{)}$ $\boxed{\text{ENTER}}$

Answer: 2.5775

5. Calculator: 3780 $\boxed{\text{log}}$

Graphing Calculator: $\boxed{\text{log}}$ $\boxed{(}$ 3780 $\boxed{)}$ $\boxed{\text{ENTER}}$

Answer: 3.5775

9. Scientific Calculator: 37,800 $\boxed{\text{log}}$

Graphing Calculator: $\boxed{\text{log}}$ $\boxed{(}$ 37,800 $\boxed{)}$ $\boxed{\text{ENTER}}$

Answer: 4.5775

13. Scientific Calculator: 2010 $\boxed{\log}$

Graphing Calculator: $\boxed{\log}$ $\boxed{(}$ 2010 $\boxed{)}$ $\boxed{\text{ENTER}}$

Answer: 3.3032

17. Scientific Calculator: 2.8802 $\boxed{10^x}$ $\left(\text{or } \boxed{\text{inv}} \ \boxed{\log}\right)$

Graphing Calculator: $\boxed{\text{2nd}}$ $\boxed{\log}$ $\boxed{(}$ 2.8802 $\boxed{)}$ $\boxed{\text{ENTER}}$

Answer: 759

21. Scientific Calculator: 3.1553 $\boxed{10^x}$

Graphing Calculator: $\boxed{\text{2nd}}$ $\boxed{\log}$ $\boxed{(}$ 3.1553 $\boxed{)}$ $\boxed{\text{ENTER}}$

Answer: 1430

25. $\text{pH} = -\log[\text{H}^+]$

$\quad = -\log\left(6.5 \times 10^{-4}\right)$

$\quad = -\log(0.00065)$

$\quad = -(-3.2)$

$\quad = 3.2$

29. $\quad M = \log T$

$\quad 5.5 = \log T$ Substitute given value

$\quad T = 316{,}000$ Use $\boxed{10^x}$ on the calculator

33. $\quad M = \log T$

$\quad 6.5 = \log T$ Substitute given value

$\quad T = 3{,}160{,}000$ Use $\boxed{10^x}$ on the calculator

Now, we compare this to our answer from #29 and find it is 10 times as large.

37. $\log(1 - r) = \dfrac{1}{t} \log \dfrac{w}{p}$

$\log(1 - r) = \dfrac{1}{5} \log \dfrac{5750}{7550}$ Substitute given values

$\log(1 - r) = \dfrac{1}{5}(-0.1183)$ Use $\boxed{\log}$ on calculator

Continued on next page.

$$\log(1 - r) = -0.0237 \qquad \text{Multiply}$$
$$1 - r = 0.947 \qquad \text{Use } \boxed{10^x} \text{ on calculator}$$
$$r = 0.053 \qquad \text{Solve linear equation}$$
$$r = 5.3\% \qquad \text{Change to a percent}$$

41. $\ln e^5 = 5 \ln e \qquad \text{Property 3}$

$\qquad = 5(1) \qquad \log_e e = 1$

$\qquad = 5 \qquad \text{Multiply}$

45. $\ln 10e^{3t} = \ln 10 + \ln e^{3t} \qquad \text{Property 1}$

$\qquad = \ln 10 + 3t \ln e \qquad \text{Property 3}$

$\qquad = \ln 10 + 3t(1) \qquad \ln e = 1$

$\qquad = \ln 10 + 3t \qquad \text{Multiply}$

49. $\ln 15 = \ln(5 \cdot 3)$

$\qquad = \ln 5 + \ln 3 \qquad \text{Property 1}$

$\qquad = 1.6094 + 1.0986 \qquad \text{Substitute given values}$

$\qquad = 2.7080 \qquad \text{Add}$

53. $\ln 9 = \ln 3^2$

$\qquad = 2 \ln 3 \qquad \text{Property 3}$

$\qquad = 2(0.10986) \qquad \text{Substitute given value}$

$\qquad = 2.1972 \qquad \text{Multiply}$

Problem Set A.5

1. $\qquad 3^x = 5$

$\log 3^x = \log 5 \qquad \text{Take log of both sides}$

$x \log 3 = \log 5 \qquad \text{Property 3}$

$x = \dfrac{\log 5}{\log 3} \qquad \text{Divide both sides by log 3}$

$x = \dfrac{0.6990}{0.4771} \qquad \text{Use } \boxed{\log} \text{ on calculator}$

$x = 1.4650 \qquad \text{Divide}$

5.
$$5^{-x} = 12$$

$\log 5^{-x} = \log 12$	Take log of both sides
$-x \log 5 = \log 12$	Property 3
$x = -\dfrac{\log 12}{\log 5}$	Divide both sides by $-\log 5$
$x = -\dfrac{1.0792}{0.6990}$	Use $\boxed{\log}$ on calculator
$= -1.5440$	Divide

9.
$$3^{2x+1} = 2$$

$\log 3^{2x+1} = \log 2$	Take log of both sides
$(2x + 1) \log 3 = \log 2$	Property 3
$2x + 1 = \dfrac{\log 2}{\log 3}$	Divide both sides by $\log 3$
$2x = \dfrac{\log 2}{\log 3} - 1$	Subtract 1 from both sides
$x = \dfrac{1}{2}\left(\dfrac{\log 2}{\log 3} - 1\right)$	Multiply both sides by $\dfrac{1}{2}$
$x = \dfrac{1}{2}\left(\dfrac{0.3010}{0.4771} - 1\right)$	Use $\boxed{\log}$ on calculator
$x = \dfrac{1}{2}(0.6309 - 1)$	Divide
$x = \dfrac{1}{2}(-0.3691)$	Subtract
$x = -0.1846$	Multiply

13.
$$A = P(1 + \frac{r}{n})^{nt} \quad \text{where} \quad P = \$500, A = \$1000, r = 0.06, n = 2$$

$1000 = 500\left(1 + \dfrac{0.06}{2}\right)^{2t}$	Substitute given values
$1000 = 500\,(1.03)^{2t}$	Simplify
$(1.03)^{2t} = 2$	Divide both sides by 500
$\log(1.03)^{2t} = \log 2$	Take log of both sides
$2t \log 1.03 = \log 2$	Property 3
$t = \dfrac{\log 2}{2 \log 1.03}$	Divide both sides by $2 \log 1.03$

$$t = \frac{0.3010}{0.0257} \qquad \text{Use } \boxed{\text{log}} \text{ on calculator}$$

$$t = 11.7 \text{ years} \qquad \text{Divide}$$

17. $\log_8 16 = \dfrac{\log 16}{\log 8}$ **21.** $\log_7 15 = \dfrac{\log 15}{\log 7}$

$$= \frac{1.2041}{0.9031} \qquad\qquad\qquad = \frac{1.1761}{0.8451}$$

$$= 1.3333 \qquad\qquad\qquad\quad = 1.3917$$

25. Scientific Calculator: $345 \boxed{\text{LN}}$

Graphing Calculator: $\boxed{\text{LN}} \boxed{(}\ 345\ \boxed{)}\ \boxed{\text{ENTER}}$

Answer: 5.8435

29. Scientific Calculator: $10 \boxed{\text{LN}}$

Graphing Calculator: $\boxed{\text{LN}} \boxed{(}\ 10\ \boxed{)}\ \boxed{\text{ENTER}}$

Answer: 2.3026

33. $32{,}000\, e^{0.05t} = 64{,}000$ \qquad Substitute given value

$\qquad\quad e^{0.05t} = 2$ \qquad\qquad\quad Divide both sides by 32,000

$\qquad\quad 0.05t = \ln 2$ \qquad\qquad\quad Change to logarithmic form

$\qquad\qquad t = \dfrac{\ln 2}{0.05}$ \qquad\qquad Divide both sides by 0.05

$\qquad\qquad t = \dfrac{0.6931}{0.05}$ \qquad\quad Use $\boxed{\text{LN}}$ on calculator

$\qquad\qquad t = 13.9$ years or toward the end of 2001

37. $A = Pe^{rt}$

$\dfrac{A}{P} = e^{rt}$ \qquad\qquad\qquad Divide both sides by P

$rt = \ln \dfrac{A}{P}$ \qquad\qquad\quad Change to logarithmic form

$t = \dfrac{1}{r} \ln \dfrac{A}{P}$ \qquad\qquad Multiply both sides by $\dfrac{1}{r}$

41.

$$A = P(1 - r)^t$$

$$\frac{A}{P} = (1 - r)^t \qquad \text{Divide both sides by } P$$

$$\log(1 - r)^t = \log \frac{A}{P} \qquad \text{Take log of both sides}$$

$$t \log(1 - r) = \log A - \log P \qquad \text{Property 3 and Property 2}$$

$$t = \frac{\log A - \log P}{\log(1 - r)} \qquad \text{Divide both sides by } \log(1 - r)$$